Steel Structures

Considerations to Reduce Failures Due to Instability

Sponsored by the
Forensic Engineering Division (FED) of the
American Society of Civil Engineers

Prepared by the
Committee on Practice to Reduce Failures

Peter J. Maranian, P.E., S.E.

Published by the American Society of Civil Engineers

Library of Congress Cataloging-in-Publication Data

Names: Maranian, Peter, author.
Title: Steel Structures: Considerations to reduce failures due to instability / by Peter J. Maranian (for ASCE FED CPRF Committee).
Description: Reston : American Society of Civil Engineers, 2021. | Includes bibliographical references and index. | Summary: "Steel Structures: Considerations to Reduce Failures due to Instability provides a detailed overview of the issues associated with the instability of steel structures"-- Provided by publisher.
Identifiers: LCCN 2021022254 | ISBN 9780784415832 (paperback) | ISBN 9780784483640 (ebook)
Subjects: LCSH: Structural failures--Prevention. | Building, Iron and steel.
Classification: LCC TA656 .M365 2021 | DDC 624.1/71--dc23
LC record available at https://lccn.loc.gov/2021022254

Published by American Society of Civil Engineers
1801 Alexander Bell Drive
Reston, Virginia 20191-4382
www.asce.org/bookstore|ascelibrary.org

***Errata:* Errata, if any, can be found at https://doi.org/10.1061/9780784415832.**

ISBN 978-0-7844-1583-2 (print)
ISBN 978-0-7844- 8364-0 (PDF)

Manufactured in the United States of America.

27 26 25 24 23 22 21 1 2 3 4 5

Cover credit: Back cover, black and white photo: courtesy of Ian Firth

Dedicated to my father and mother, Maurice and Angela Maranian,
for their courage through many adversities including during WWI and WWII.

Contents

Preface

During the development of structures in the last two centuries, there has been continuing progress toward both slender and less heavy structures.

In the building and bridge industries, structures have developed from masonry to timber to cast iron and then to steel and concrete, further advanced by prestressed concrete.

As structures have developed to utilize slender elements, the phenomenon of buckling instability, a precarious form of failure, has increasingly been encountered.

This publication primarily focuses on the instability of steel structures including the various forms of instability associated with compressive forces, bending, shear, and torsion, including global and local buckling.

There have been numerous failures because of instability both during erection and service that have occurred over more than a century. These have led to the tragic consequences of the failure of many buildings and bridges. Although the subject of instability is quite well understood by researchers and engineers, failures because of instability still occur.

Chapter 1: Examples of failures that have taken place during the nineteenth and twentieth centuries, intended to illustrate the instability problems that can occur because of a lack of understanding of the issues.

Chapter 2: Brief history and an introduction to the theory of instability. This is intended to describe some of the extensive research and development that has taken place in two centuries.

Chapter 3: Aspects of material considerations along with details and the practical aspects associated with various components of steel construction.

Chapter 4: Various load demands along with possible long-term deterioration issues.

Chapter 5: Discussion and recommendations regarding current practice both for temporary conditions and for permanent design to limit failures of steel structures because of instability.

This book is intended to provide an overview of the issues of the potential instability of steel structures and hopefully improve the understanding of structural and civil engineers and others, including architects, contractors, steel fabricators, and erectors. It is also intended to illustrate the need for diligence with regard to consideration for the stability of steel structures during service and construction, which cannot be overestimated. There have been many failures of structures during service and construction that have occurred because of a lack of appropriate care with respect to stability.

This book is not intended to replace authoritative publications on the subject of instability that have been produced to date including publications such as Zeimian (2010). Reference to other, more in-depth publications on instability for further reading is provided. Although references to some codes and specifications are mentioned, the intent is to, by way of historical overview, improve the understanding of and the respect for the various forms of instability that can occur.

The book may not include the requirements of current codes and specifications on related items as the emphasis is based more on a historical overview.

Application of measures to mitigate instability should be carried out by adopting the minimum standards of the applicable code with a thorough understanding of the subject.

This manuscript has been peer-reviewed for technical accuracy, with significant advice and recommendations by Ashwani Dhalwala, S.E., the former chair of ASCE Forensic Engineering Division (FED), Committee on Practice to Reduce Failures (CPRF).

The manuscript was also reviewed by ASCE FED's Executive Committee, including Ben Cornelius and Stewart Verhurst, along with a review by members of CPRF, Vince Campisi, P.E., M.ASCE (chair), William J. Schmitz, P.E. (Ret.), and Theodor Francu, S.E.

Acknowledgments

Following several, always lively meetings, which took place circa 2010 to 2012, for the Committee of Practice to Reduce Failures (CPRF) [one of the committees of ASCE's Forensic Engineering Division (FED)], I volunteered to embark on drafting this publication, with the support of the committee, with the intent that it may help reduce the reoccurrence of failures of steel structures because of instability. The document has taken several years to compile and could not have been carried out without the significant support, work, consultation, review, and advice of many for whom I offer my wholehearted thanks.

During the early stages of drafting the document, the late Ruben Zallen, who was a highly regarded member of FED/CPRF for many years, gave me much good advice regarding the critically important historical publications on the subject of instability. As the document progressed, I constantly received much encouragement and advice from members of CPRF and FED's Executive Committee. This included good guidance from Alicia Diaz De Leon (past chair, FED) and Dr. Navid Nastar (past chair, FED) and the benefit of critical review with valuable comments from Ben Cornelius (chair, FED), Stewart Verhulst (FED Executive Committee member), Vince Campisi (CPRF chair), William Schmitz (CPRF member) and Theodor Francu (CPRF member). Throughout the somewhat lengthy time that the publication evolved, I have had the fortune and privilege of receiving much good advice and recommendations from Ashwani Dhalwala (past chair, CPRF) derived from his exceptional knowledge and experience in structural engineering. On completion of the final draft of the publication, Ashwani carried out a peer review for technical accuracy, as well as contributed to the publication writing the important section on nonlinear buckling (Section 2.13). I am also thankful to Ian Firth from the United Kingdom, who provided me with highly valuable photographs of steel box girder failures.

I am very thankful to have had the help of Sema Akyurek, who carried out all the sketches and figures, along with Beth Aldrich, who did much of the typing-both of whom tolerated my endless revisions through to its final stages. Tricia Bovey, acquisitions editor, along with Michie Gluck, books production manager with ASCE have been so very helpful and diligent in facilitating to process the manuscript toward its publication.

Finally, I wish to thank my wife, Srpuhi, for her many years of patience and understanding while I delved into the research and writing for this publication.

CHAPTER 1
Case Histories

1.1 FAILURES DURING CONSTRUCTION

- **Aircraft hanger, Nigeria, 1978**: *Collapse of a hanger* (circa 1979). The one-story structure was constructed of metal deck on steel "Z" purlins on a series of approximately 60 m (200 ft) long trusses. A substantial portion of the roof collapsed when installing the last truss, resulting in one fatality. This was found to be because of erectors pulling the bottom chord of the truss to connect bracing trusses. The top chord was not yet connected to the bracing truss, and the purlins had not been installed (Figure 1-1a, b). Also, the cranes, which had installed the truss, had been released. This resulted in the lack of stability of the last truss, with the top chord of the truss in compression, leading to its failure because it could not be arrested because the cranes had been released. As the truss failed, it progressively brought down adjacent trusses to the extent that about half of the roof effectively collapsed.

  The sequence should have been such as to safeguard the truss by maintaining the cranes in operation until bracing trusses and sufficient purlins had been installed to maintain stability.

- **28-Story building, Los Angeles, California, 1985**: *Eleven stories of a 28-story structure in Los Angeles* collapsed during erection circa 1985 when steel beams and columns, yet to be erected, were temporarily placed on erected metal deck and steel framing. Steel members, many of them heavy sections, were being loaded onto the fifth floor by a derrick when a section of the floor, about 40 ft × 50 ft (12.1 m × 15.2 m), gave way, causing partial progressive failure of the remaining 10 floors below and collapsing all the way down to the basement floor. Figure 1-2(a) shows the construction prior to the accident and shows the derrick and steel members temporarily stacked on the deck and steel framing already installed. Figures 1-2(b and c) shows the floor plan and the area where the collapse occurred, respectively. Three ironworkers tragically perished in the accident, and others were badly injured. Wood sleepers had been placed on the metal deck to support the steel members. However, these were not properly aligned with the beams, such that the loads from the temporarily stacked beams caused bending of the metal deck and

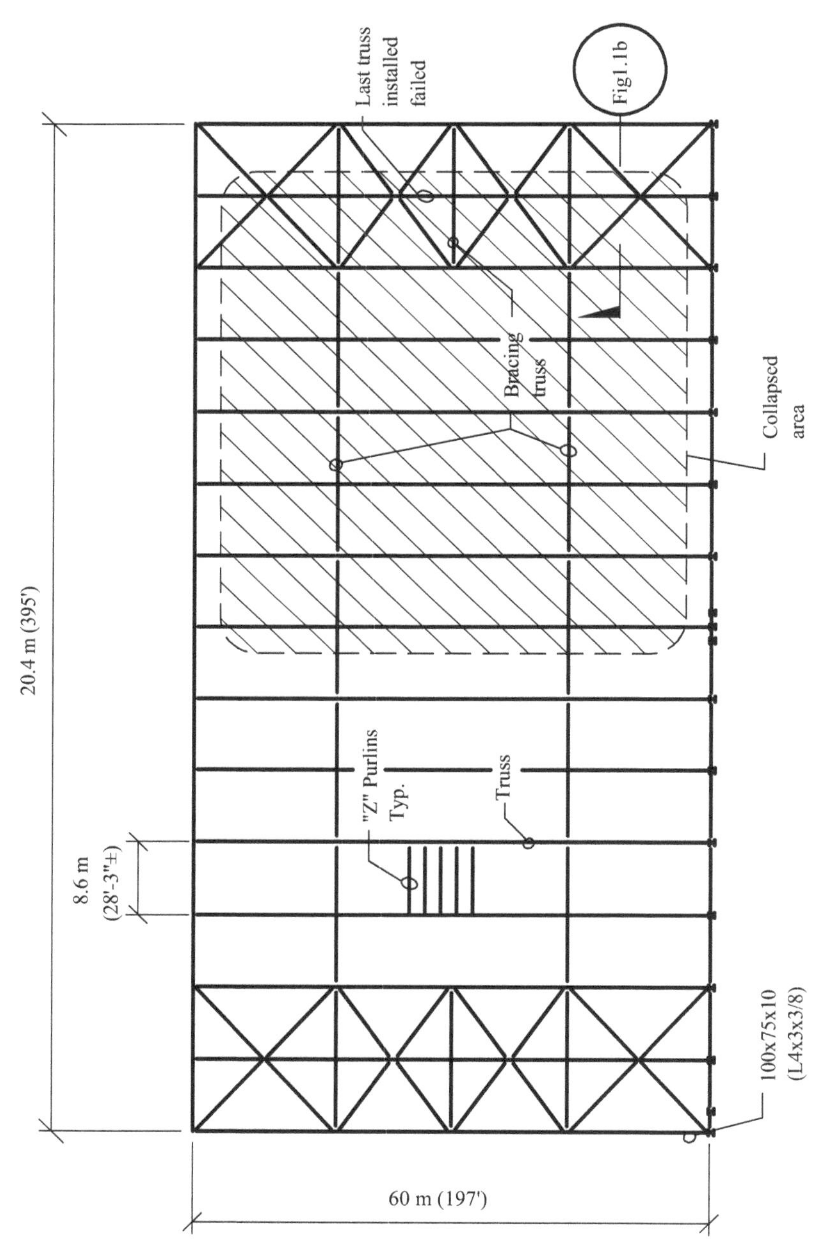

Figure 1-1.(a) Aircraft hanger in Nigeria. Partial collapse during construction (1978): Roof plan.

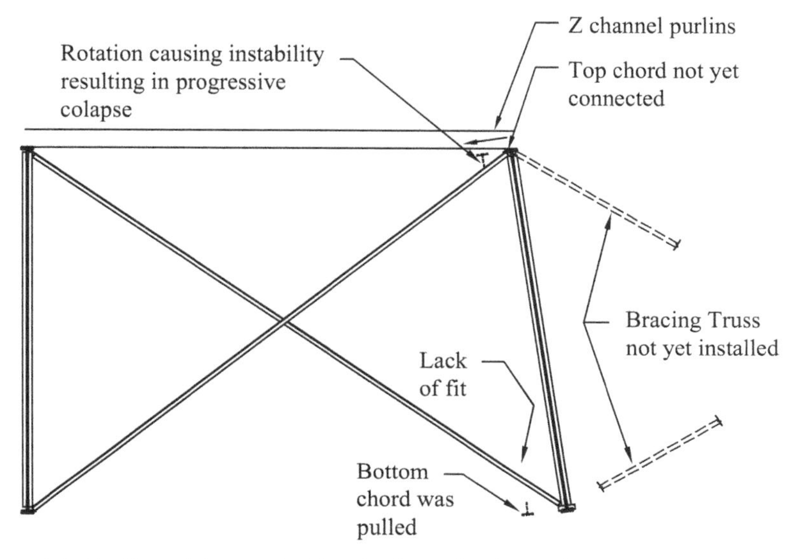

Figure 1-1.(b) Aircraft hanger in Nigeria. Partial collapse during construction (1978): Section.

eccentric loading of the beams. Furthermore, the steel metal deck had been placed but only tack-welded. Figure 1-2(d) shows the derrick and a partial area of the collapse. Figure 1-2(e) shows the area of the collapse. The framing also comprised beams framing into columns with tee-shaped steel-seated connections and a steel angle connecting the top beam flange to the web of the column, as shown in Figure 1-2(f). However, these top angles had not yet been welded to the column web. Figure 1-2(g) shows the end of a fallen beam with a partially buckled web and a top angle, which was required to be welded to a column, with no evidence of a weld. The misalignment of the wood sleepers, along with the lack of completion of deck welding and lack of top angles at the beam to column-seated connections, in the opinion of the writer, allowed instability to occur when the temporary loading was placed on the steel framing. When the steel members were placed, the loads were applied eccentrically to at least one of the beams. This caused twisting of the beams, creating the potential for instability. The metal deck, because it was only tack-welded to the steel beams, did not appear to offer adequate restraint. Furthermore, because the top angles were not welded and the beams were only supported by seated connections, the twisting action on the beams could not readily be restrained. The rotation of the beams at the columns caused the floor to give way, taking along with it the beams framing

Figure 1-2.(a) 28-story building, Los Angeles, California. Partial collapse during construction (1985): Photograph of construction prior to accident.

to the girders, the metal deck, the wood sleepers, the temporary steel, and the three ironworkers.

The deck welding and welding of the top angles at the beam to column-seated connections should have been carried out along with a proper alignment of wood sleepers. These measures, to ensure stability, would have prevented the collapse that tragically led to three fatalities.

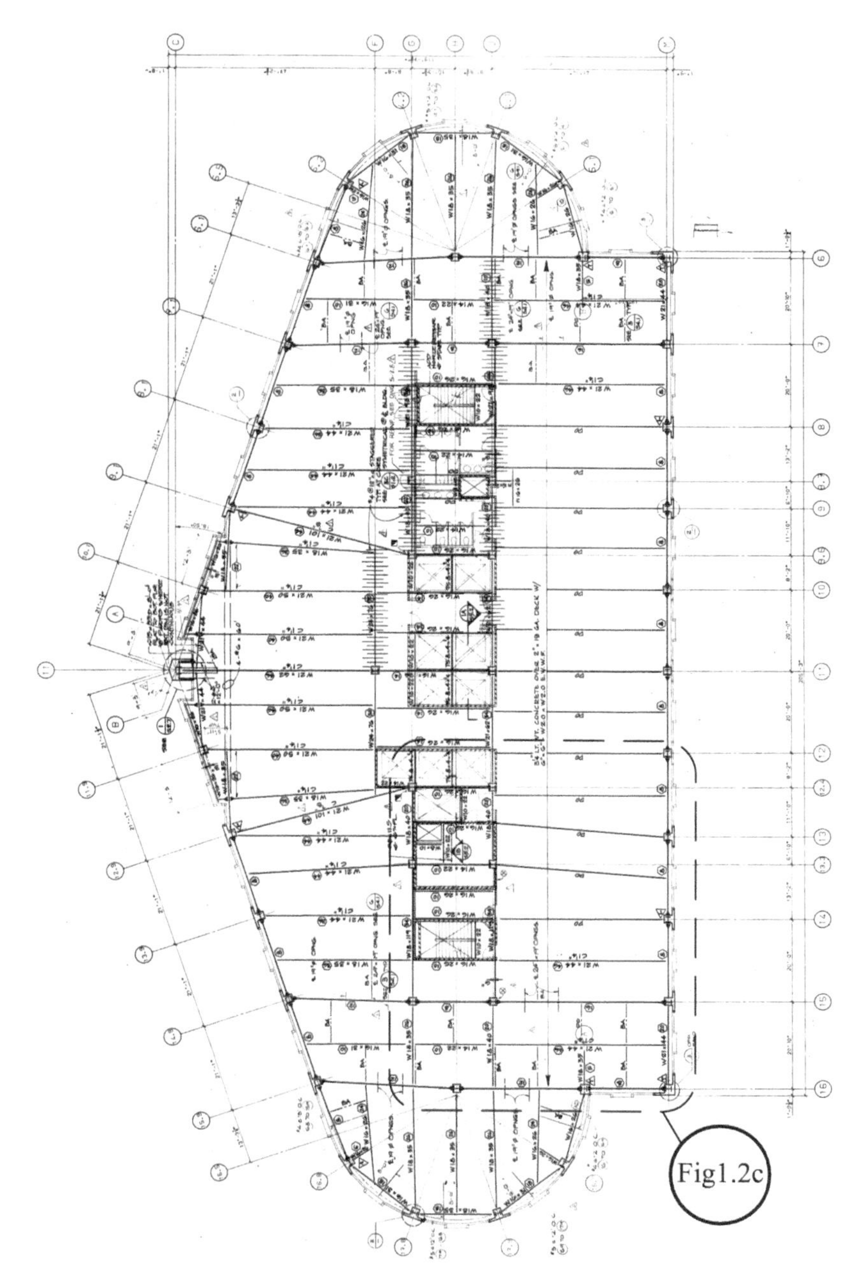

Figure 1-2.(b) 28-story building, Los Angeles, California. Partial collapse during construction (1985): Typical floor framing plan.

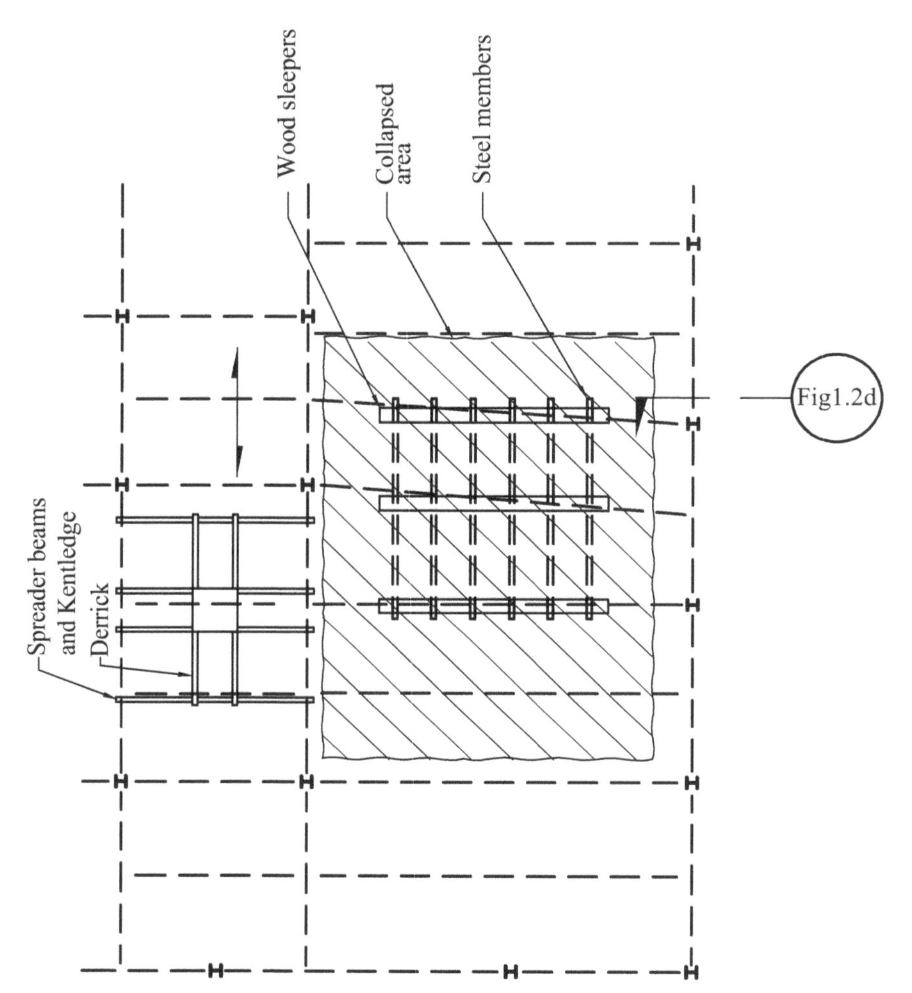

Figure 1-2.(c) 28-story building, Los Angeles, California. Partial collapse during construction (1985): Partial third floor plan.

- **Middle School Complex, New England, 1996**: Deep long-span steel joists collapsed during the construction of a high-bay section, injuring three ironworkers (Zallen 2003). Only seven of the 110 ft 0 in (33.52 m). span joists had been installed when the erection foreperson realized that some of the joists were not in the right place.

  To remove the joists, cross-bridging (shown in Figure 1-3) had to be removed. During the removal, the already erected slender joists buckled and collapsed. With reference to Figure 1-3, the first two joists (LS-1 and LS-2)

Figure 1-2.(d) 28-story building, Los Angeles, California. Partial collapse during construction (1985): View to east looking at bottom of hole at areas of collapse.

Figure 1-2.(e) 28-story building, Los Angeles, California. Partial collapse during construction (1985): View looking to south and west showing area of collapse and derricks.

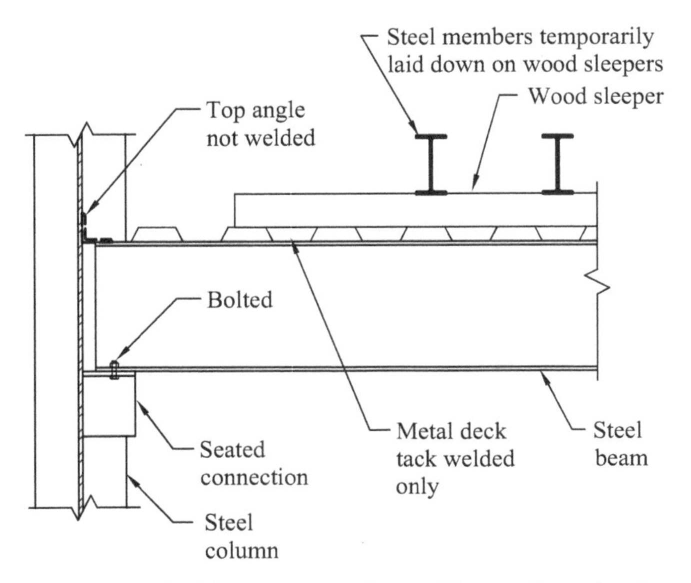

Figure 1-2.(f) 28-story building, Los Angeles, California. Partial collapse during construction (1985): Section at beam connection.

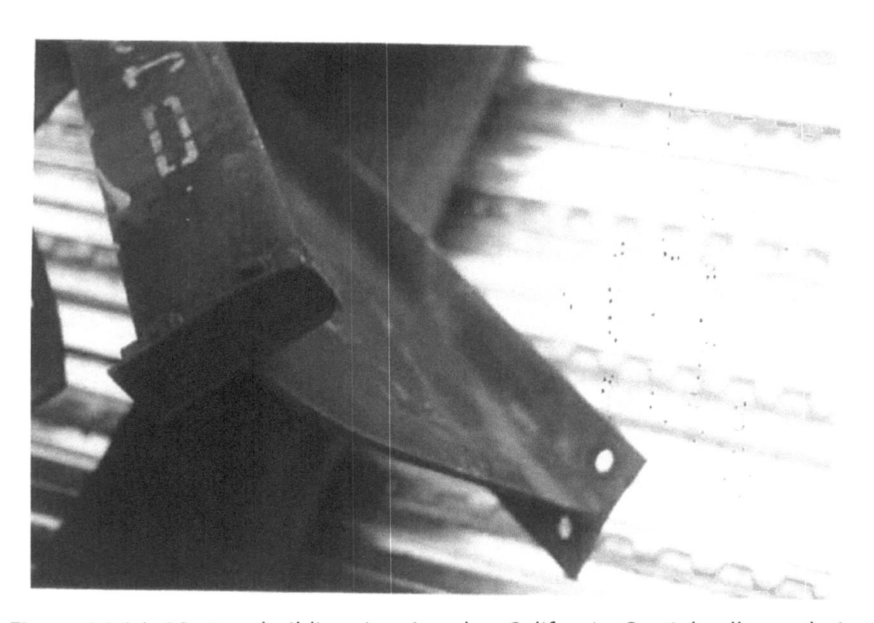

Figure 1-2.(g) 28-story building, Los Angeles, California. Partial collapse during construction (1985): End of beam that was connected to column.

were erected together with cross-bridging and strong backs, followed by the remaining joists. Five rows of cross-bridging were spaced at approximately 20 ft at the center. The last three joists LS-S, LS-6, and LS-7 were close together so that they only had horizontal bridging.

The cross bracing at the three center bays was first substantially, but not completely, removed. When this occurred, joists LS-5, LS-6, and LS-7 buckled laterally to the right. The lateral displacement of these joists caused the remaining cross bracing to pull the four other joists, causing them to buckle and collapse.

Zallen (2003) was critical of the standard procedures given in Technical Digest No. 9 saying that, even though the erector did not follow all the requirements, anchoring the ends of the cross-bridging to a line of beams and columns, as required by Technical Digest No. 9, would not have prevented the collapse.

Zallen was of the opinion that there was a lack of an adequate factor of safety for stability. Careful consideration of the bracing system needs to be taken into account to ensure that it has adequate stiffness. Horizontal angle struts should have been installed and/or steel roof decking should have been installed immediately after the steel framework had been plumbed and all bolts had been tightened. This would have increased the stiffness of the bracing system, affording stability and thus preventing the collapse.

- **Milford Haven Bridge Collapse, Wales, 1970**: The steel box girder bridge over the River Cleddau in Wales, on June 2, 1970, collapsed during construction, killing four workers and injuring five others [Figure 1-4(a to c)]. The collapse occurred during the launching of the single continuous welded box girder, which had no substantial stiffening except at the piers. The cause of the failure was complex but was thought to be because of the buckling of the stiffened diaphragm over a pier. The stiffened diaphragm, for some unknown reasons, was reduced from .75 in to .375 in. (19 mm to 9.5 m) in thickness and became inadequate to accommodate for the significant demands imposed by cantilevering during launching. The buckled diaphragm subsequently initiated buckling of the adjacent flanges and webs (Macleod 2007, Firth 2010, Brady 2016, Bridle and Sims 2009).
- **Westgate Bridge, Australia, 1970**: Soon after the Milford Haven Bridge disaster, the Westgate Bridge over the Yarra River in Melbourne, Australia, collapsed in October 1970, killing 35 workers. The bridge comprised a trapezoidal box, with long transverse cantilevers on each side of the box section, as shown in Figure 1-5(a) The collapse occurred between piers 10 and 11, which spans 367 ft (112 m) with long transverse cantilevers at each side of the box section [Figure 1-5(a)]. The trapezoidal box was erected about 164 ft (50 m) above ground in two halves, commencing with the north half, followed by the south half. When first erected and the temporary trestles were removed, a buckle suddenly developed and the north half was found

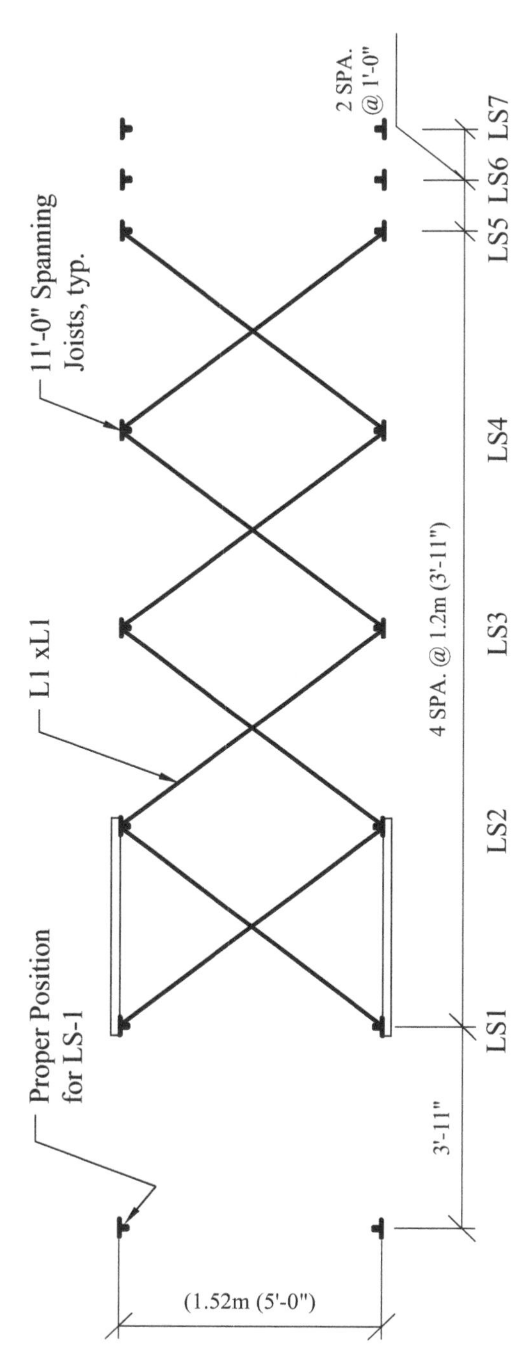

Figure 1-3. Schematic section through erected joists.
Source: Reproduced from Zahlen (2003), courtesy of ASCE.

Figure 1-4.(a) Milford Haven (Cleddau) Bridge, Wales collapse (1970): Showing substantial buckling of box girder.

Source: Courtesy of Ian Firth, with permission from COWI UK.

to be 15 in. (380 mm) higher than the south half, partly because of the asymmetry of the two sections and the long cantilevers at each side of the box. The transverse stiffeners were inadequate to restrain the longitudinal stiffeners. Also, the splice detail at the longitudinal stiffeners, which occurred every approximately 52 ft (16 m), involved a gap of about 12.5 in. (318 mm) with a single plate, smaller in area than the longitudinal stiffener, offset such that it caused eccentric forces to occur. In lieu of lowering the two halves to the ground, remedial measures were carried out in place to rectify the difference in elevation between the two halves. These included

Figure 1-4.(b) Milford Haven (Cleddau) Bridge, Wales collapse (1970): Showing collapsed span.

removing some bolts in the transverse stiffeners at the south half. These procedures resulted in complications as deflections of each half were still found to be different by about 4.5 in. (114 mm). To address the deflection differential, it was decided to ballast the high side with ten 8 ton blocks. This caused some local buckling and slipping of plates. Bolts were loosened and/or removed to try to address this issue. Initially, the buckling was partially flattened out. However, appreciable movements were occurring, and a significant buckle of the inner web took place. A gentle settlement of the bridge occurred primarily because of yielding and possibly thermal changes. Attempts to rebolt the connection between the transverse diaphragm including enlarging holes and the upper flange plate were made. This required a diagonal brace to be removed. The increased instability resulted in the collapse of the bridge. Figure 1-5(b) indicates the stages of the collapse. Poor communication and a lack of proper supervision and support from the designers were thought to have contributed to the disaster. A Royal Commission inquiry by the Government of the State of Victoria, Australia, which included additional research on thin steel plate behavior, was carried out to implement the changes to design and construction (Firth 2010, APEA 1990, Brady 2016).

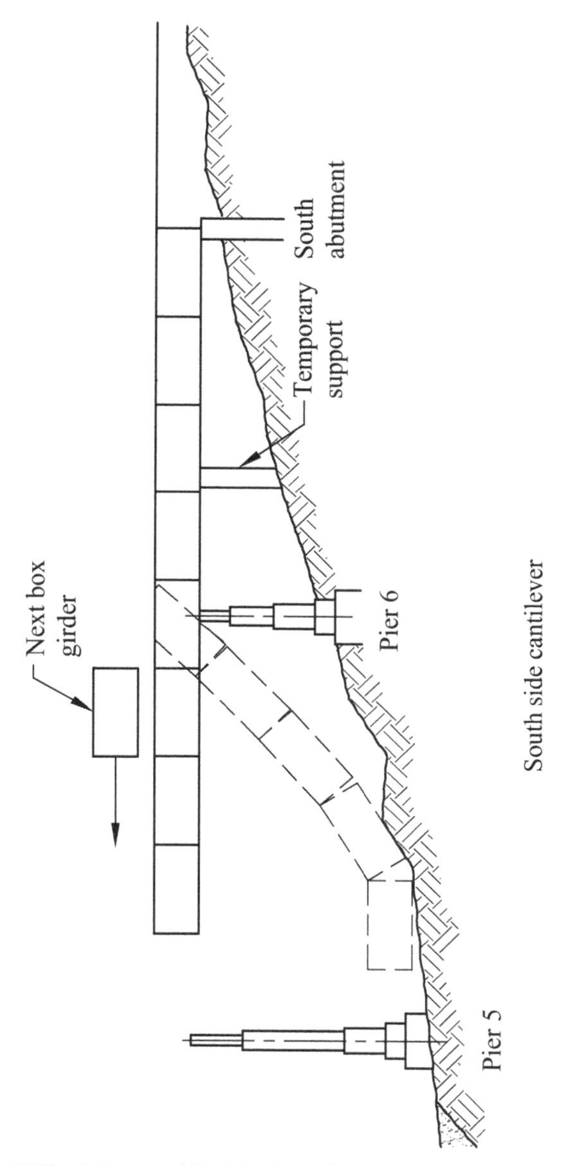

Figure 1-4.(c) Milford Haven (Cleddau) Bridge, Wales collapse (1970): Showing construction stage at time of collapse.

- **Koblenz Bridge, Germany, 1971**: The Koblenz Bridge over the River Rhine collapse during construction on November 10, 1971, was apparently caused by a combination of reasons (Figure 1-6a, b). The steel box girders had high temporary bending moments during the construction stages when they performed as significant cantilevers. A failure occurred at a box girder

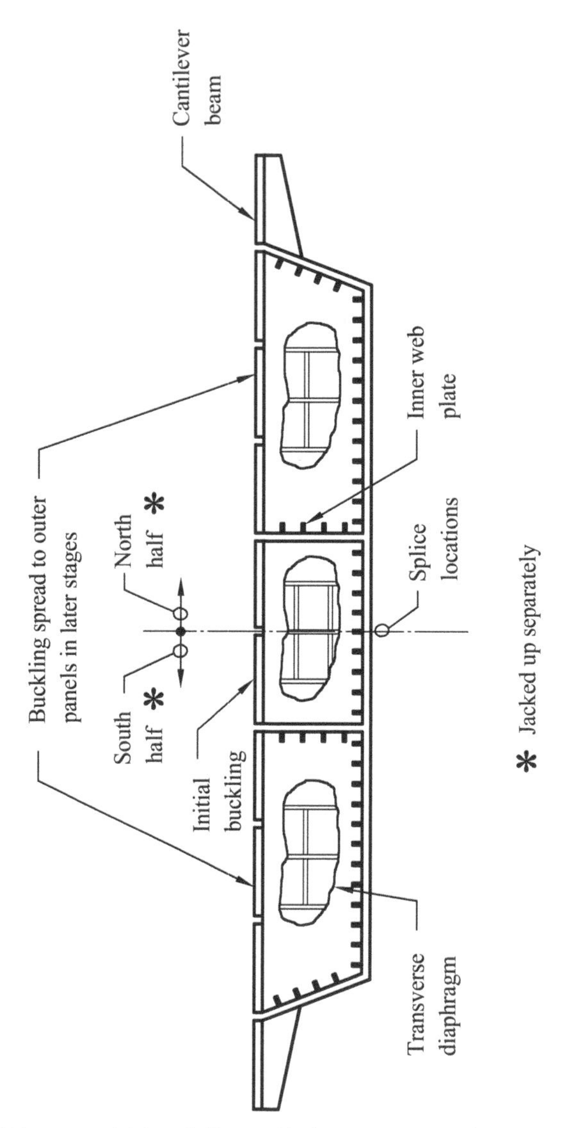

Figure 1-5.(a) Westgate Bridge Collapse Melbourne, Australia, 1970: Section.

splice joint. The complete penetration weld joining the flange plates caused significant distortion, increasing the eccentricity of the compressive forces being applied to the plates. Detailing was poor, such that there was also a gap of approximately 15 in (381 m). in the connection of vertical stiffeners to the flanges of the box girders, such that the flanges were locally unstiffened. It was thought that this may have occurred because of the stiffeners being cut back at the bottom plate to allow for access for welding. Apparently, the missing

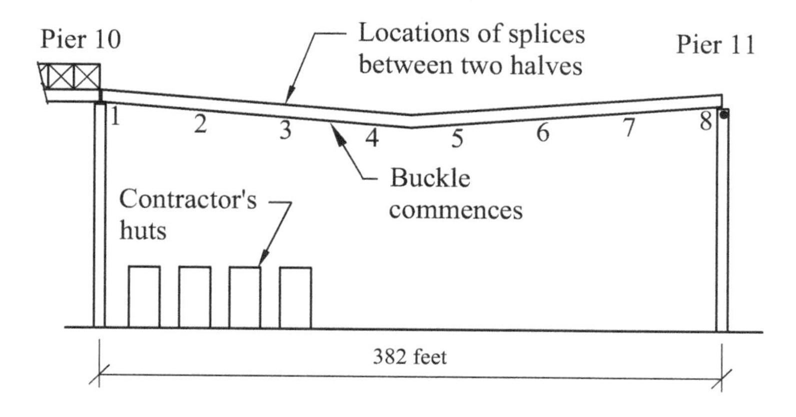

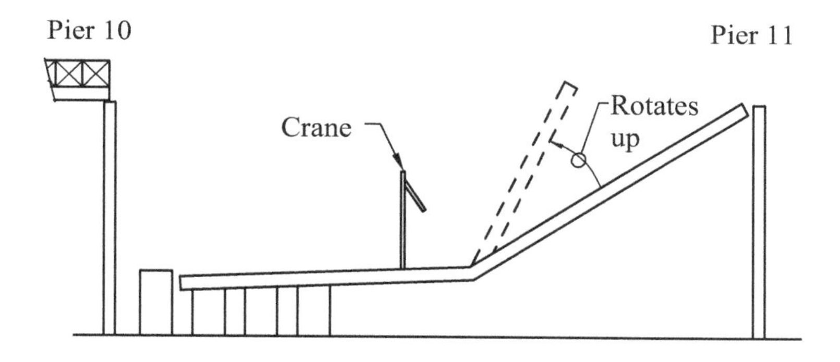

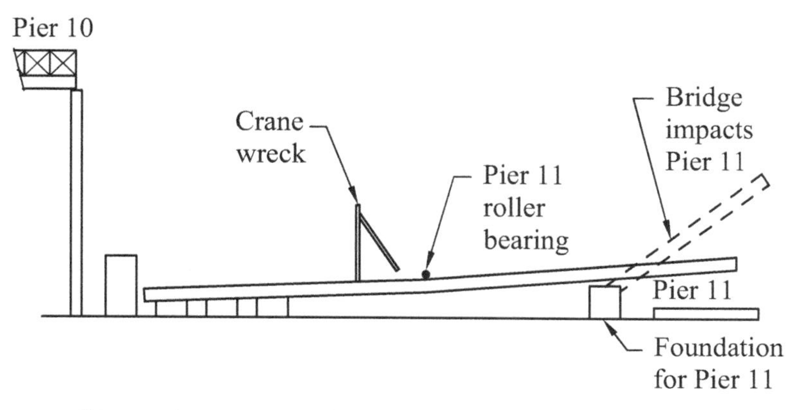

Figure 1-5.(b) Initial, interim, and final collapse stages.

Figure 1-6.(a) Koblenz Bridge collapse, Germany (1971).
Source: Courtesy of Ian Firth, with permission from COWI UK.

stiffener plates were replaced but were welded only to the vertical stiffeners and not to the bottom plate. The unstiffened length was actually effectively greater than 15 in (.38 m). This led to local buckling, as the cantilever was extended, increasing the load on the bottom plate, thus causing a collapse of the bridge (Firth 2010, Brady 2016).

- **Merrison Report on Box Girder Bridges**: The collapses of the Milford Haven Bridge and Yarra Bridge instigated the UK government to set up investigations by a committee of experts led by Dr. Merrison from Bristol University, England. The basis of the investigations included examining the design rules and providing recommendations for further research. It was found that, although much knowledge on analysis had been developed, it was not in a form readily available to practicing engineers. The initial investigations led to "Interim Rules" being issued in May 1971. Subsequent extensive research

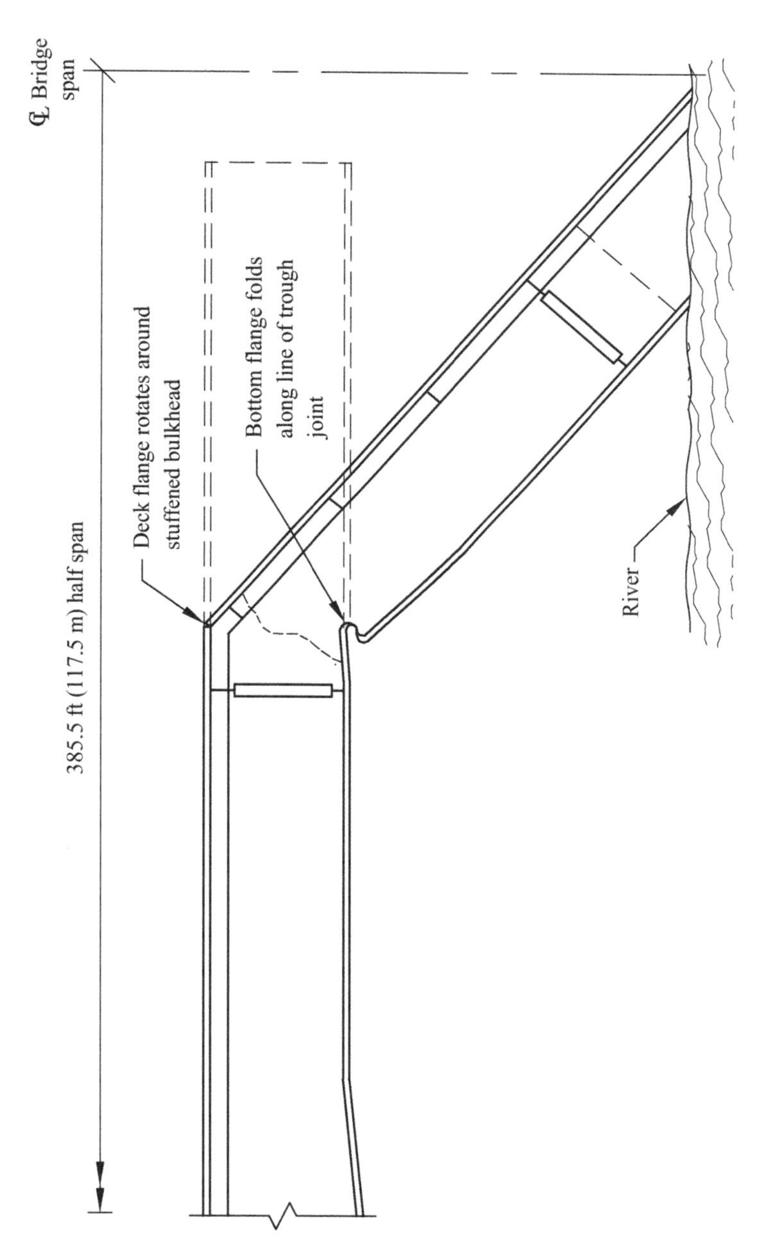

Figure 1-6.(b) Koblenz Bridge collapse, Germany (1971): Partial elevation.

Source: Reproduced from Bridle and Sims (2009), courtesy of the Institute of Civil Engineers, United Kingdom.

led to publication of a document titled, "Interim Design and Workmanship Rules" (Merrison 1973). These rules included requirements for the stress analysis of box girders and for their connections in complex stress fields. The rules also included the effects of residual stresses and geometric imperfections associated with fabrication tolerances. A consideration of secondary stresses, associated with residual stresses, along with as-built eccentricities (both those built-in and those arising from fabrication tolerances and weld distortion) were found to be of significant importance in design. The Merrison Committee also recommended (1) an independent check on the design, (2) independent check on the erection and design of temporary works, (3) a clear delineation of responsibility between the engineer and the contractor, and (4) provision by the engineer and contractor of adequately qualified supervisory staff on-site (Firth 2010, Bridle and Sims 2009).

1.2 FAILURES DURING SERVICE

- Several failures of open wrought iron bridges, with "pony trusses," occurred in Western Europe and Russia in the late nineteenth century (Timoshenko 1953, p. 296). Insufficient lateral rigidity of the upper chord in compression led to the failures. A lack of understanding of the flexural rigidities of the top chord and vertical members exists. Figure 1-7, sketched from a figure in Timoshenko (1953, p. 296), illustrates the typical failures that occurred.

Figure 1-7. Failure of an open bridge (3D view).

Source: Derived from Figure 181 in Timoshenko (1953, p. 296).

- **Knickerbocker Theater Collapse, Washington, DC, 1922:** The theater building structure, housing 1,800 seats, collapsed during the evening of January 28, 1922, when approximately 2.5 ft (.76 m) of snow had collected on the roof from a storm that had started on January 23, 1922. The temperature was around 20 °F (−5 °C) at the time of the collapse. The collapse occurred as the orchestra commenced with an overture. The north wall, adjacent to the stage, first cracked open, quickly followed by the roof and parts of the balcony collapsing. The bad weather had kept many theatergoers away, but there were still approximately 400 people in the theater. Tragically, 98 people were killed and another 133 were injured.

  The building, built in 1916, utilized steel roof trusses that were substituted for plate girders by the contractor but were found, subsequent to the collapse, to be underdesigned. The snow loading was estimated to be about 75 lb/ft^2 (366.1 kg/m^2), substantially greater than the design load of 23 lb/ft^2 (112.2 kg/m^2). The main truss had been reduced by 9 in (228.6 m). in depth in the redesign, causing it to deflect significantly. Trusses were also not anchored to the walls, and the effect of successive expansion during the summers and contraction during the winters was thought to have caused the north wall to progressively move outward. During the event on January 28, 1922, the main truss deflected significantly, shortening the top chord and lengthening the bottom chord. The investigators thought that the shortening of the main truss top chord may have been such that it caused the loss of support at the bearings. Also, the main exterior walls below the main truss may have been pushed outward as it bowed and as a result unseated the truss itself. However, the investigators also cited a lack of adequate bracing at the connection of the main truss and column. It was thought that the bearings became overloaded, which caused eccentricities that resulted in buckling of the main truss. All three factors were thought to have contributed to the disaster (Schlager 1994).

- **Hartford Civic Center Coliseum Roof Collapse, Hartford, Connecticut, 1978:** During the early hours of January 18, 1978, the roof of the approximately 300 ft × 360 ft (91.4 m × 109.72 m) space frame roof structure of the 10,000-seat facility collapsed 83 ft (25.3 m) to the floor only a few hours after about 5,000 people had occupied the facility to watch a college basketball game. Snow had accumulated on the roof weighing as much as 19 lb/ft^2 (92.7 kg/m^2) on the roof of the structure. The structure had been completed just a few years earlier, in 1973.

  The design of the structure, carried out circa 1971, utilized an innovative concept intended to reduce costs compared with a more standard space frame roof design. The roof space frame structure, supported on four columns, comprised steel members at the upper and lower grids every 30 ft (9.15 m) with steel diagonal steel members every 30 ft [Figure 1-8(a to c)]. Additional steel members were provided at the upper horizontal members, and the diagonal members were braced at their midpoints. Truss members were made up of four steel angles welded back to back, providing a cruciform section that has

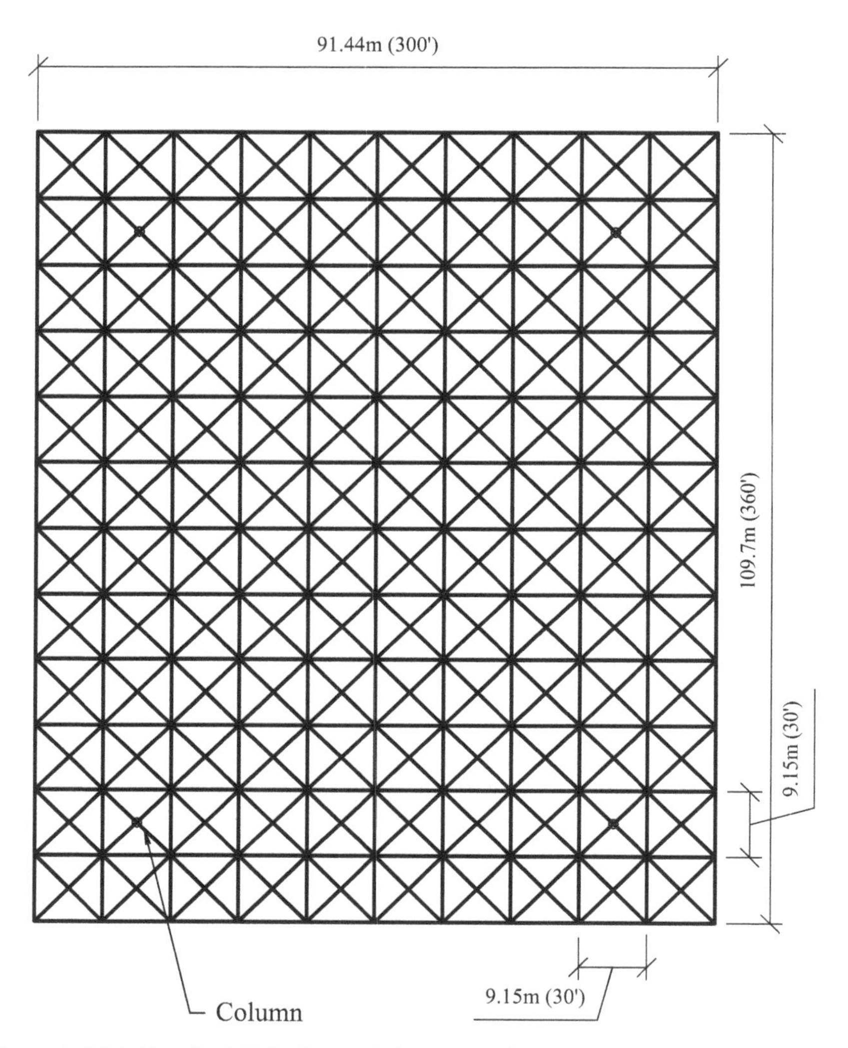

Figure 1-8.(a) Hartford Civic Center Coliseum roof collapse (1978): Framing plan.
Source: Reproduced from Lev Zetlin Associates, courtesy of Thornton Tomasetti.

a significantly lower buckling resistance compared with "I" beams and tube sections. Short posts, typically located at nodes, supported the roof panels. It had been reported during construction that deflection of the space frame, which was not cambered, measured 12 in. (304.8 mm) compared with 8.5 in., estimated by computer analysis. However, no corrective action was taken during construction. Also, wide flange members, supporting the roof, were themselves supported by stub columns and, thus, were not sufficiently effective in bracing the top chord of the trusses (Kaminetzky 1991, Levy and Salvadori 1992).

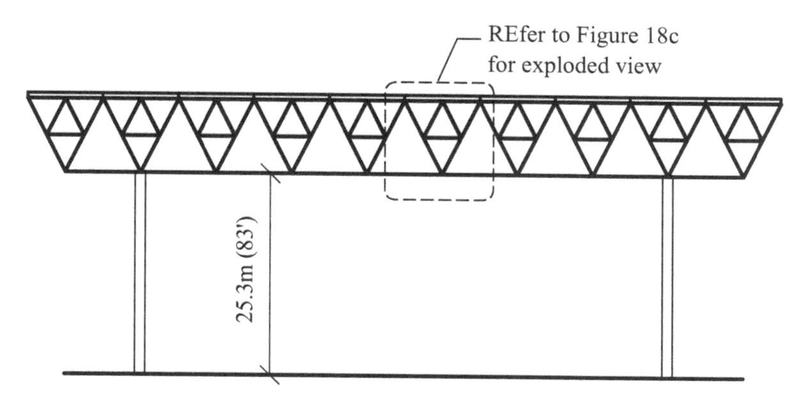

Figure 1-8.(b) Hartford Civic Center Coliseum roof collapse (1978): Section.

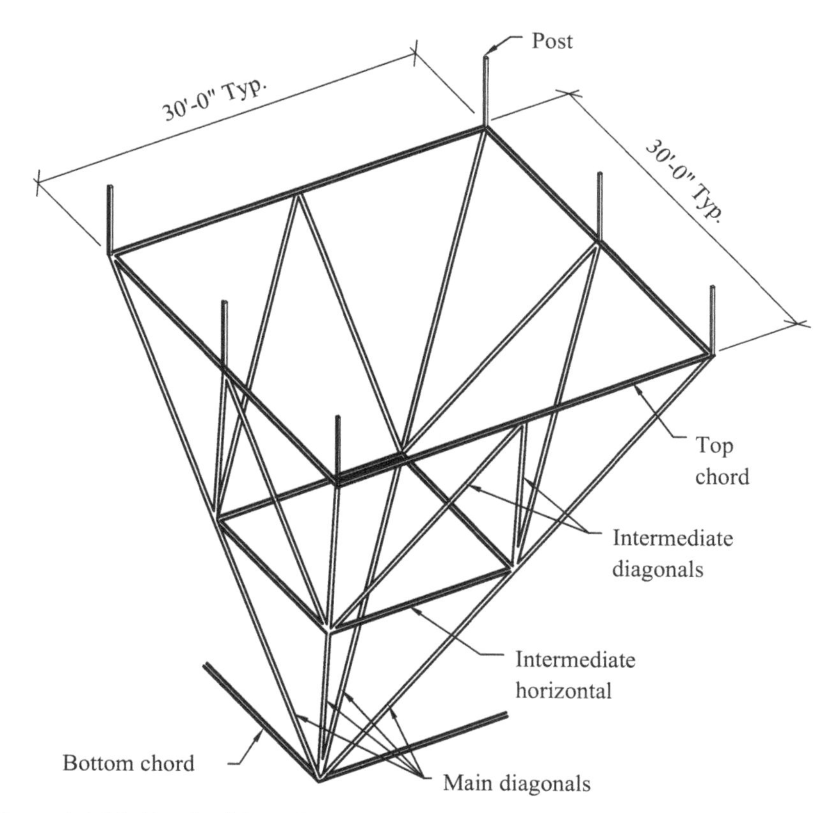

Figure 1-8.(c) Hartford Civic Center Coliseum roof collapse (1978): 3D view.

One investigator was of the opinion that the structure was underdesigned and should have been designed for greater loading. Another investigator thought that the failure of a weld, supporting the scoreboard weighing 75 tons (68038.9 kg), caused the collapse. Yet, other investigators were of the opinion that diagonal members were subjected to torsional buckling, resulting in a twisting of the truss members leading to the collapse. It was also found that as-built connection details between the top horizontal members and the diagonal members differed significantly from the original design, resulting in appreciably greater eccentricities than accounted for in the design. The increased eccentricities reduced the ability of the diagonal members to brace the top chord. Other issues included exterior members being braced only every 30 ft (9.15 m) instead of 15 ft (4.57 m) as specified in the design and posts landing on top chord members away from nodes, thus inducing significant bending stresses (Levy and Salvadori 1992, Kaminetsky 1991, Delatte 2009). Delatte (2009) cited factors including the underestimation of the dead load, the overloading of exterior compression members as much as 852%, and the overloading of interior members by as much as 72%. These findings were indicative of how a lack of adequate bracing can significantly affect stability.

- **Partial Failure of a Single-Story Market, Burbank, California, 1993:** A single-story building built in 1961, sustained a partial failure of a major spine member subjected to large compressive forces. The building, which was used as a market, was 177 ft (54 m) by 156 ft (47.6 m) in the plan and utilized a steel cable catenary roof structure. The building comprised of steel bulb tees and a gypsum roof deck, supported by a series of catenary (approximately parabolic) cables, typically at 5 ft (1.5 m) at the center, which were connected to steel trusses, slightly inclined to the horizontal, and in turn, supported on concrete block walls. The roof effectively spanned 156 ft (47.6 m). The forces on steel trusses, which provided anchorage for the cables, were resisted by braced frames at the ends and by the central spine member.

  The central spine member comprised a W36 × 300, with its web horizontal, connecting to two diagonal W36 × 260, forming a "Y" at each side. These W36 × 300 members, which were orientated sideways (flanges vertical), were, in turn, connected to the steel trusses [Figure 1-9(a to d)]. A pipe column occurred at the center of the W36 × 300 spine member. This W36 × 300 spine member was required to resist a substantial approximately horizontal compressive force because of the forces of tension in the cables.

  In early May 1993, a contractor, carrying out remodeling, noticed that the ceiling of the market had lowered in one area by about 2 ft (.6 m). On inspection, the central W36 × 300 spine member was found to have buckled and yielded at one location, causing it to deflect significantly downward. It also showed signs of twisting. The downward deflection was fortunately arrested by two sets of cables.

  Previous concerns for the deflection by structural engineering consultants were reported circa 1978. The remodeling work, carried out in 1993, about the

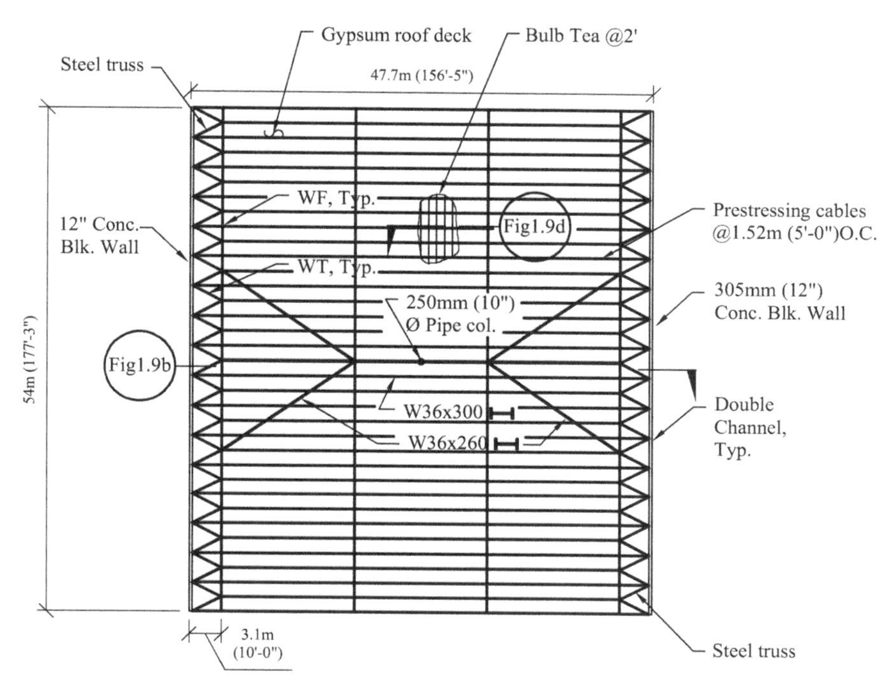

Figure 1-9.(a) Single-story market Burbank, California (1993): Roof plan.

time of the failure, included mechanical changes and new openings through the roof. A pipe column at one of the intersections was temporarily removed and reinstalled during the remodeling.

In the opinion of the investigators, the main W36× spine member and two diagonal W36 × 300 members were thought to have a bow. Also, the tension forces in the cables, immediately below the main spine member, may have relaxed. This relaxation of the cables potentially caused further deflection, increasing the instability of the member. It is possible that the minor modifications carried out in the remodeling induced additional changes that caused the instability and substantial yielding. However, in the opinion of the writer, who was one of the investigators, stability considerations of the W36× members were inadequate in the original design. Insufficient support members were present to the main spine member to ensure adequate stability. Also, the cable anchorages at the steel trusses were eccentric to the plane of the truss, which caused secondary stresses further adding to the instability effects [Figure 1-9(c)].

The retrofit, to ensure stability, involved adding a series of posts along the main spine member. The building continued to operate for a few years before it was eventually demolished and replaced.

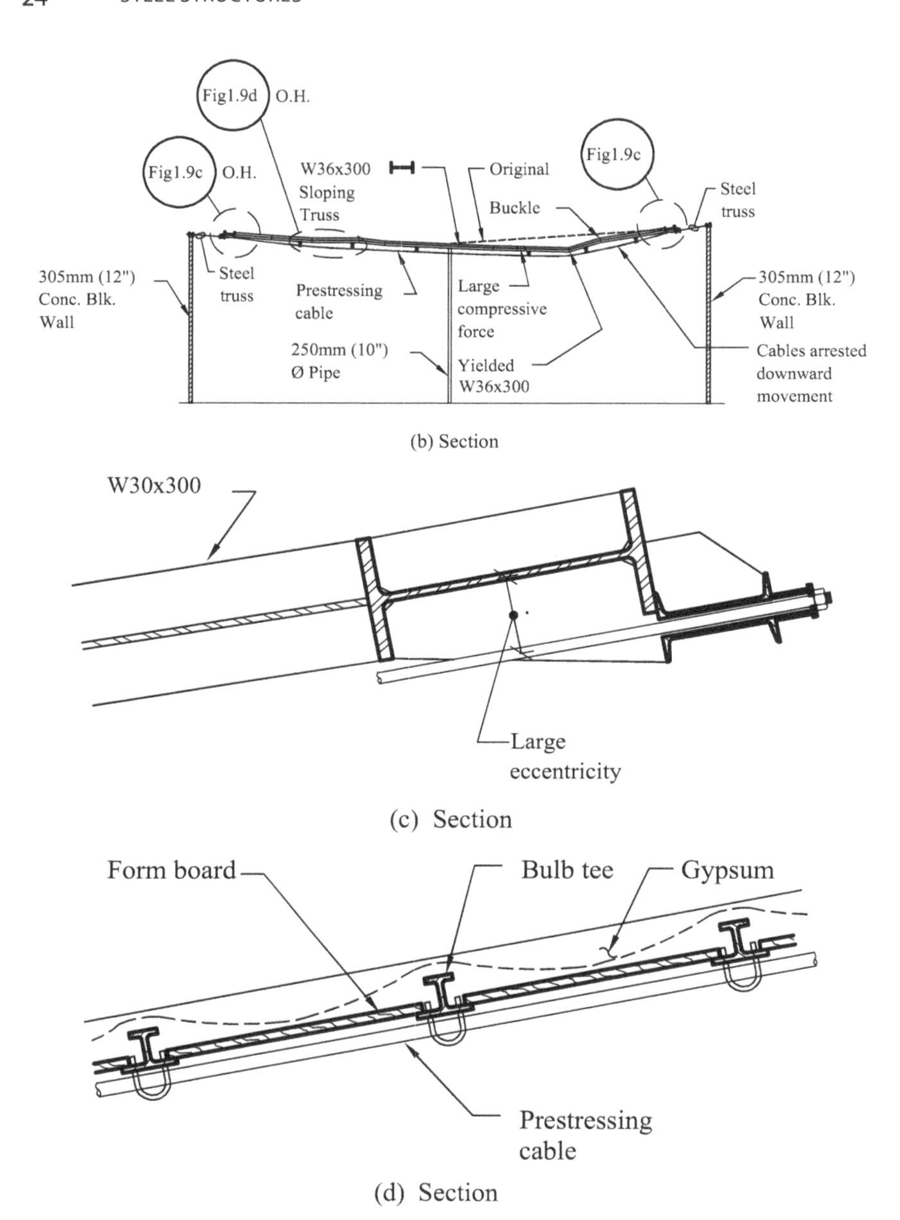

Figure 1-9.(b through d) Single-story market, Burbank, California (1993): Sections.

- **Cantilever failure in a Residence, Southern California, 2010:** One of three cantilevered steel tapered girders, supporting a two-story wood-framed house overhanging the beach adjacent to the Pacific Ocean, failed during a major storm. A portion of the house also collapsed.

Figure 1-10.(a) Partial collapse of residence in Southern California (2010).

Figure 1-10.(b) Partial collapse of residence in Southern California (2010): Close up showing cantilever failure and severe corrosion.

The steel girder, which cantilevered about 40 ft (12.1 m), was tapered in depth. The width of the flanges was also reduced along the cantilever. The flanges were welded to the web.

As can be seen from Figure 1-10(a, b), severe buckling of the web occurred where the flanges bent along with separation of the bottom flange of the girder from the web occurring about 10 ft (3 m) from the end of the cantilever. The unpainted girder had suffered severe corrosion indicated by flaking and pitting such that a thickness of only .06 in. (1.5 mm) was left at some locations.

The significant corrosion appreciably weakened the beam such that, along with only minimal lateral bracing, the girder was inadequate to sustain the building loads and the dynamic forces from the sea waves, generated during the storm.

This example illustrates how corrosion can not only significantly reduce the strength of members but can also increase instability, primarily because of the reduction in thicknesses.

References

APEA (Association of Professional Engineers). 1990. *Collapse of the west gate bridge, occupational health and safety for engineers*. South Melbourne, AU: APEA.

Brady, S. 2016. "Westgate bridge collapse—The story of the box-girder bridge." *Struct. Eng.* 94 (10): 26–28.

Bridle, R. J., and F. A. Sims. 2009. "The effect of bridge failures on UK technical policy and practice." *Eng. Hist. Heritage* 162: E1–E11.

Delatte, N. J. 2009. *Beyond failure, forensic studies for civil engineers*. Reston, VA: ASCE.

Firth, I. 2010. "Lessons from history—The steel box girder story." *Struct. Eng.* 88 (5): 2.

Kaminetzky, D. 1991. *Design and construction failures: Lessons from forensic investigations*. New York: McGraw-Hill.

Levy, M., and M. Salvadori. 1992. *Why buildings fall down*. New York: W.W. Norton, 68–75.

Macleod, I. A. 2007. "Structural engineering competence in the computer era." *Struct. Eng.* 85 (3): 35–39.

Merrison, A. W. 1973. *Inquiry into the basis of design and method of erection of steel box-girder bridges*. Great Britain Department of Environment; Great Britain Scottish Development; Great Britain Welsh Office. London: Her Majesty's Stationary Office.

Schlager, N. 1994. *When technology fails*. Detroit: Gale Research.

Timoshenko, S. P. 1953. *History of strength of materials*. New York: Dover.

Zallen, R. M. 2003. "Problems with industry standard for erecting open web steel joists." In *Proc., 3rd Forensic Engineering Congress*, edited by P. A. Bosela, N. J. Delatte, and K. L. Rens. Reston, VA: ASCE, 477–484.

CHAPTER 2

Theory of Instability Including Historical Development

2.1 INTRODUCTION

In researching the nearly 300 year history of testing and development on buckling and instability, the writer found it intriguing to learn how long theories have taken to be established.

The well-known Euler theory, on the elastic behavior of axially loaded members, took about 100 years to be validated and accepted. This was followed by approximately another 100 years to establish reasonable agreement for axially loaded struts with lower slenderness ratios, where yielding at the extreme fibers occurs.

Much testing and development has taken place in the last 300 years, such that there is very good understanding of the issues associated with buckling. However, there still remain some issues that have led to appreciable differences in the buckling provisions given in various codes (i.e., Europe, United States, Canada.)

In recent times, some global and local buckling, beyond yielding, has been permitted in seismic resisting systems (e.g., steel moment frames, special concentric braced frames, steel-plated shear walls). However, buckling when the material is beyond yield is the least quantifiable buckling mode.

In the writer's opinion, at least a limited knowledge of the historical background of the development of the buckling theory and associated investigations appears important to help in the understanding of the complex nature of buckling. The following overview intends to provide a historical overview of the subject of buckling and instability. It should be understood that the following discussion includes only representative formulas to illustrate the development of buckling theory.

2.2 AXIAL LOAD INSTABILITY

The first known testing of wooden rectangular struts was carried out by Petrus van Musschenbroek circa 1729 in Germany (refer to Figure 2-1, which is based on

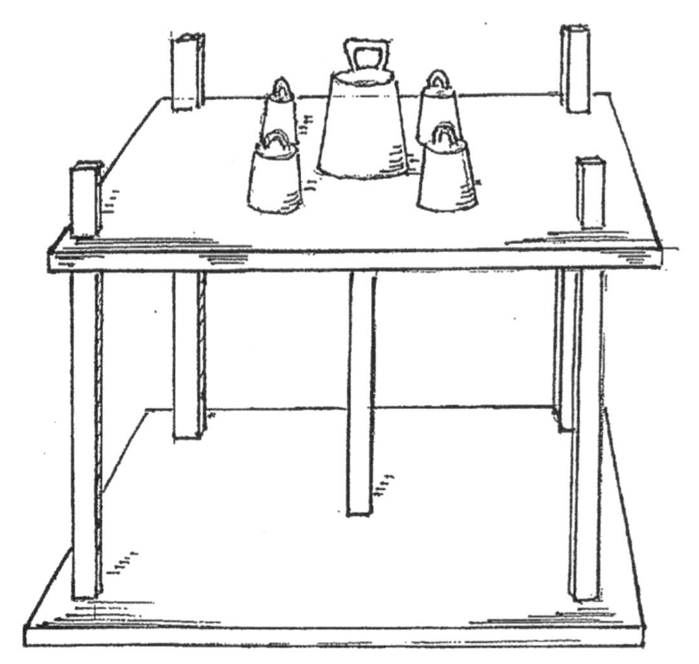

Figure 2-1. Musschenbroek's apparatus for compression tests.
Source: Reproduced from Figure 40 in Timoshenko (1953, p. 56), courtesy of Dover.

a figure in Timoshenko 1953, p. 56). Musschenbroek concluded that the buckling load is inversely proportional to the square of the length of the strut, which was later mathematically shown by Leonard Euler (Timoshenko 1953, pp. 55–56).

The first major theoretical work on buckling was conducted by Leonard Euler (1707 to 1783). Euler was born in Basel, Germany, subsequently settled in St. Petersburg, Russia, and then later returned to Berlin. Following the establishment and understanding of elastic behavior in the late seventeenth century by R. Hooke, leading to Hooke's law, Euler, an outstanding mathematician, investigated the elastic behavior of elastic bars using variational calculus. This also followed Jacob Bernoulli's earlier investigations and theoretical development on prismatic bars. Euler's work led to his recognition that the compressive load on columns was limited due to buckling. In 1744, Euler established that the load was inversely proportional to the height of the column such that a column twice as high would only have a quarter of the load capacity. Eventually, Euler published his work in 1757 and gave a simplified formula for the critical load, P, as follows (Timoshenko 1953, p. 34):

$$P = \frac{C\pi^2}{4L^2} \tag{2-1}$$

where C is a constant, and L is the height of the column.

The production of wrought iron–rolled shapes commenced in 1783 and began to be used in buildings and structures. This led to an increased interest in verifying the column theory.

A. J. L. Lagrange (1736 to 1813), born in Turin, Italy, also investigated elastic curves and established an equation for columns about 1770, similar to that derived by Euler as follows (Timoshenko 1953, p. 38):

$$C = \frac{d^2y}{dx^2} = -Py \tag{2-2}$$

Lagrange also showed that an infinite number of buckling curves were possible.

This is represented when a small axial load, P, is centrally applied to a perfectly straight column; it remains straight and is only subject to axial compression, as shown in Figure 2-2. If a small lateral force, H, is also applied simultaneously with the axial load, a small deflection occurs. On release of these forces, the column returns to its straight form. If the axial load, P, is increased, eventually its magnitude reaches the critical load, P_{cr}, when only a small horizontal force is

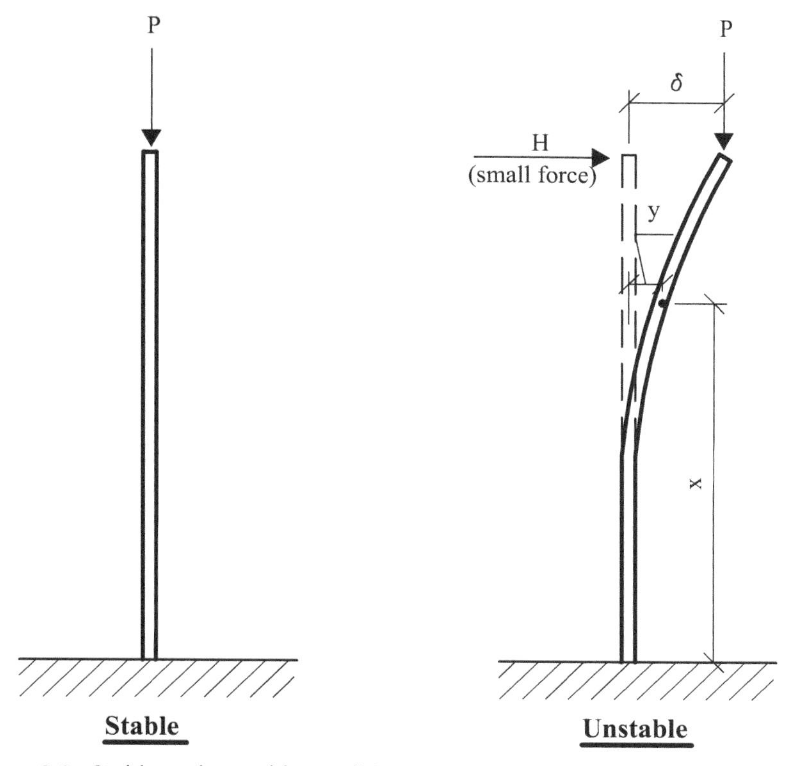

Figure 2-2. Stable and unstable conditions.

sufficient to cause instability again, as illustrated in Figure 2-2. This instability is represented by the following equations (Timoshenko 1953, pp. 28–36; Popov 1959, pp. 345–346; Timoshenko and Gere 1961, pp. 46–49):

$$M = P(\delta - y)$$

$$EI\frac{d^2y}{d^2x^2} = P(\delta - y)$$

where
M = Moment,
E = Modulus of elasticity or Young's modulus, and
I = Moment of inertia.

For δ, x, and y, see Figure 2-2.

The case was solved by Euler and improved by Lagrange, as previously stated, based on the deflection curve's being a half sine wave. The general relationship is as follows (Timoshenko and Gere 1961, p. 49):

$$P_{cr} = \frac{\pi^2 EI}{L^2} \tag{2-3}$$

Note that Euler used the notation "*C*" that, in the nineteenth century, eventually became *EI*.

It should be understood that Equations (2-1) to (2-3) apply to perfectly elastic columns and do not consider any potential for yielding, which is discussed subsequently. At the P_{cr} load, the deflected form, in a half sine wave, is retained and with any increase in load, the perfectly elastic column becomes unstable.

If the load is only slightly increased above P_{cr}, a significant lateral deflection occurs, as shown in Figure 2-3. According to Popov (1952, p. 343), an increase of only 1.5% causes a lateral deflection of 22% of the length of the column. Gerard (1962) states for an L/r [known as *slenderness ratio*, defined with Equation (2-10)] of 100, that if the axial load is increased by only 1%, the maximum fiber stresses become 880% of the average axial stress.

Thomas Young (1773 to 1829), born in Somerset, England, also made contributions to the column buckling theory. Young was both a doctor and a scientist and is best known for establishing the principles of deformation that led to the term modulus of elasticity, also known as Young's modulus, E. In 1807, Young established a formula to determine the deflection of the curve because of a column based on an initial deflection (Timoshenko 1953, pp. 90–98).

Thomas Young considered a prismatic column with an initial slight curvature represented by one-half wave of the sine curve $= \delta_0 \sin = (\pi x L)$ (Figure 2-4). Young found the deflection at midheight to be as follows (Timoshenko 1953, p. 95):

$$\delta = \frac{\delta_0}{(1 - (PL^2 / EI\pi^2))} \tag{2-4}$$

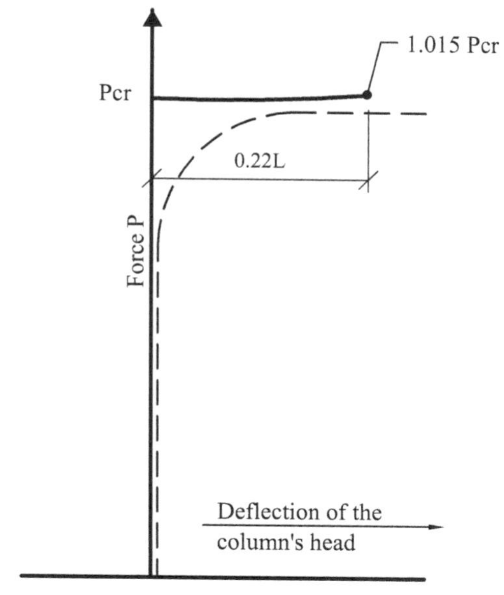

Figure 2-3. Deflection of the column's head.

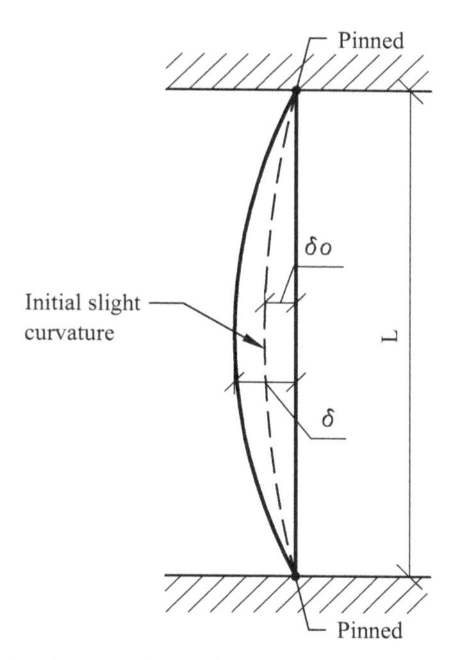

Figure 2-4. Prismatic column with initial slight curvature.

It can be seen that when the load $P = EI\pi^2/L^2$, the deflection can become infinite and, thus, unstable, as had been demonstrated by Euler and Lagrange.

Young also established the case of a slender cantilevered column with an axial force P applied eccentrically, as shown in Figure 2-5 (derived from Timoshenko 1953, p. 95). Young showed the deflection, y, to be represented by the following equation:

$$y = e\frac{(1 - \cos px)}{\cos pL} \tag{2-5}$$

where $p = \sqrt{P/EI}$.

For y, x, and e, see Figure 2-5.

During the latter part of the eighteenth century and early part of the nineteenth century, investigators in France, J. R. Perronet, then J. E. Lamblardie, followed by P. S. Girard conducted tests to try to replicate Euler's formula. The tests carried out by Girard were conducted on wooden struts and gave results that were appreciably different from perfectly elastic, thus showing no agreement with Euler's theory. Investigations were carried out by A. Duleau in 1820 at the École Polytechnique in Paris, France, and E. Hodgkinson in 1840 at the University College, London, England. Duleau used very slender bars that demonstrated good agreement with Hodgkinson's tests, which were carried out on both slender and shorter struts. The slender, solid struts showed good accord

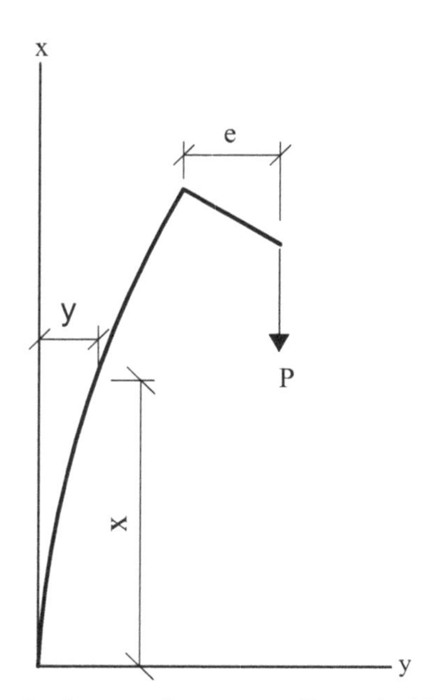

Figure 2-5. Cantilevered column with eccentrically applied load.

with Euler's formula. A satisfactory explanation by E. Lamarle, from Belgium, in 1845, showed that Euler's formula agreed with the experiments so long as the columns performed perfectly elastically and ideal end conditions were met (Timoshenko 1953).

According to Bleich (1952), it was not until 1889, more than a century since Euler first published his work, that both Considère in France and Freidrich Engesser in Germany independently formally proclaimed the validity of Euler's formula.

Further work by the British researcher R. V. Southwell, circa 1932, taking into account small initial imperfections, confirmed that Euler's formula was within engineering accuracy [less than 3% (Gerald 1962)]. Southwell's work used very careful tests on mild steel columns carried out by the German researcher Theodore von Kármán in 1910 (Timoshenko 1953, Gerald 1962).

After there was acceptance of Euler's theory, greater emphasis was made to establish theories for short columns. However, it took another century to establish theories pertaining to short columns where the material can yield (inelastic).

Lamarle, while confirming Euler's equations, also established the limits of Euler's equation. As the slenderness ratio decreases, the column is able to reach yield at its extreme fiber. The slenderness ratio at which this occurs is called the proportional limit. Ever-decreasing slenderness ratios permit a greater extent of yielding extending from the outer fibers. Thus, Equations (2-1) to (2-3), applicable to a perfectly elastic column, no longer apply (Timoshenko 1953, p. 208; Bleich 1952, p. 2). Lamarle, according to Bleich (1952), although recognizing that yielding occurs at lower slenderness ratios, did not have an insight into the behavior of columns once the material went beyond the elastic limit.

In the early part of the nineteenth century, design rules for cast-iron columns were established by the English researcher Thomas Tredgold (1788 to 1829), who was previously a carpenter. In his publication "A Practical Essay on the Strength of Cast Iron," dated 1822, Tredgold developed a series of equations that included the following equation (Timoshenko 1953, p. 209):

$$\text{Max stress, } \sigma_{\max} = \frac{P}{bh}\left(1 + \frac{aL^2}{h^2}\right) \tag{2-6}$$

where

P = Axial compressive force,
a = Certain constant,
b and h = Cross-sectional dimensions of the column, and
L = Column height.

The English engineer Lewis Gordon, using the results of experiments carried out by E. Hodgkinson and also formulas proposed by Tredgold, obtained, in the latter part of the nineteenth century, the following formula for a wrought iron rectangular section with sides $b \times h$ (Timoshenko 1953, p. 209):

$$\sigma_{\text{ult}} = \frac{36{,}000}{\left(1 + \frac{1}{12{,}000}\frac{L^2}{h^2}\right)} \tag{2-7}$$

where h is the short dimension of the column, and L is the column height.

The Scottish Researcher W. J. Macquorn Rankine, circa 1865, in applying Gordon's formula, modified it to apply to wrought iron I beams. Rankine recommended, for the ultimate compressive stress, the following (Timoshenko 1953, p. 210):

$$\sigma_{\text{ult}} = \frac{36{,}000}{\left(1 + \frac{1}{3{,}000}\frac{s^2}{t^2}\right)} \tag{2-8}$$

where t is the web thickness, and s is the distance between two adjacent stiffeners along a line at 45 degrees to the horizontal.

These semiempirical procedures, along with Euler's formula, became of increasingly more practical importance, particularly with the construction of steel railway bridges. Also, the production of steel, using the Bessemer process, started to replace wrought iron commencing in 1856. These formulas were used for the rest of the nineteenth century and beyond, despite a more rational method for short struts, proposed by the German Engineer, Hermann Scheffler (published in 1858). Scheffler's method considered eccentricity of load and also slight initial curvature just as Thomas Young had applied in his formula developed more than 50 years previously (Timoshenko 1953, p. 210).

The semiempirical formula, known as the Rankine–Gordon formula, for the allowable compressive stress, σ, and established by Rankine in the 1860s, is as follows:

$$\sigma_{\text{max}} = \frac{P}{A} + M/S \tag{2-9}$$

where

M = Moment,

S = Section modulus, and

A = Area.

The allowable average compressive stress becomes (Popov 1959, p. 357)

$$\frac{P}{A} = \frac{\sigma_{\text{max}}}{(1 + \phi(L/r)^2)} \tag{2-10}$$

where ϕ is a constant, and r is the radius of gyration $\left(\sqrt{I/A}\right)$.

Johann Bauschinger, at the Polytechnic Institute of Munich, circa 1887, carried out probably the first reliable tests on both slender and short columns

(Timoshenko 1953, p. 294). Bauschinger used conical attachments to provide free rotation at the ends. Bauschinger tests confirmed Euler's formula for slender columns, but not enough tests were carried out to establish formulas for short columns. Tests were also carried out by Ludwig von Tetmajer at the Zurich Polytechnical Institute around the 1890s. Tetmajer recommended the following formula for columns with $L/r < 110$ (Timoshenko and Gere 1961, p. 196):

$$\sigma_{cr} = 48{,}000 - 210\left(\frac{L}{r}\right) \tag{2-11}$$

The English researchers, W. E. Ayrton and J. Perry, in their paper published in 1886, also considering a slight initial curvature in the form of a half sine curve, established the following basic equation known as the Ayrton–Perry formula.

$$(\sigma_y - \sigma)(\sigma_E - \sigma) = \frac{(Aw_o\sigma\sigma_E)}{S} \tag{2-12}$$

where

σ_y = Yield stress,
σ = Maximum normal stress,
σ_E = Euler critical stress,
A = Cross-section area,
S = Section modulus, and
W_o = Imperfection, in a half sine curve.

Friedrich Engesser, a German structural engineer who contributed much to the subject during the period when there was a significant development in the theory of structures, established an inelastic theory in 1889 using a variable tangent modulus applied to Euler's equation to account for failure loads occurring at loads lower than those derived from Euler's formula. Engesser's approach was to replace Young's modulus with the tangent modulus, E_t, to represent the stress–strain relationship beyond the point of yield, known as the *proportional limit.* Considère questioned the assumption for tangent modulus made by Engesser. As a consequence, using a concept developed by Considère, Engesser made modifications to the tangent-modulus approach in 1895, presenting the reduced-modulus theory (sometimes referred to as the double modulus theory) (Timoshenko 1953, pp. 297–299; Gerald 1962). Engesser's general solution took into account the elastic and inelastic properties of the material. The tangent modulus represented a lower bound solution. The experimental studies carried out by von Kármán on mild steel columns in 1910 confirmed the reduced modulus theory applied to short columns, established by Engesser in 1895, to be acceptable. Bleich (1952, p. 3) was of the opinion that the developments by Considère and Engesser, along with von Kárman's experimental work, "maybe considered milestones in the long history of the buckling problem."

Coincidentally, in 1912, the English researcher, Southwell, presented the double modulus theory (also known as the *reduced modulus theory*), apparently without being aware of the previous work by Engesser and von Kármán.

The Russian professor and researcher F. S. Jasinsky, who was much involved in Bridge Engineering, circa 1890s, using the results of the tests by Bauschinger, Tetmajer, and Considére, published a table giving the critical compressive stress related to a range of slenderness ratios. These were much used in Russia, superseding Rankine's formula (Timoshenko 1953, pp. 195–298).

The effects of residual stresses, occurring in hot-rolled steel columns, were recognized in 1908 following axially loaded tests at the Waterdown Arsenal reported by J. E. Howard (Johnston 1976, p. 50). It was recognized that the residual stresses, because of the cooling of hot-rolled sections, were likely the cause of reduced column strengths in the columns tested.

A. Robertson, the English researcher, carrying out tests on struts, and again taking into account the slight initial curvature, further developed the Ayrton–Perry formula and approach in the 1920s. Robertson established the following formula (Steel Designer's Manual 1972, p. 881):

$$K_2P_c = \frac{Y_s+(n+1)C_0}{2}\left(\left(\frac{Y_s+(n+1)C_0}{2}\right)^2 - Y_sC_0\right)^{1/2} \tag{2-13}$$

where

P_c = Allowable average stress,
K_2 = Load (safety) factor, usually assumed to be about 1.7,
Y_s = Minimum yield stress,
C_o = Euler critical stress,
$n = 0.3\ (L/100r)^2$, and
L/r = Slenderness ratio.

This is known as the Perry–Robertson formula, which was adopted in the British Standard on the use of structural steel (BSI 1969, *Steel Designer's Manual* 1972).

A series of very accurate tests were carried out at the Berlin–Dahlem materials testing laboratory in Germany, with results first published in 1926. The Berlin–Dahlem tests, using steel material with a yield stress of 45 ksi, showed good agreement with Euler's formula for $L/r > 80$, where the columns remained elastic (Timoshenko and Gere 1961, p. 188). The Berlin–Dahlem tests showed that, for short columns, the critical stress was closer to the yield stress.

ASCE's Special Committee on Steel Column Research in 1926, using special roller-bearing blocks to obtain ends with hinged conditions, tested large H-section columns, with an average yield stress of 38.5 ksi, applying eccentric loads (Timoshenko and Gere 1961, p. 188). High axial loads were required such that they could not be applied on knife edges. One series of tests were carried out such that eccentricities were in the plane of the web. Another series of tests were carried out such that eccentricities were perpendicular to the plane of the web. The tests identified where yielding was taking place, again around an L/r of 80, and demonstrated that higher loads for eccentricities in the plane of the web compared were perpendicular to the web. Extensive tests were carried out by M. Ros in Zurich, Germany, in 1926; B. Johnston and I. Chaney at Lehigh University in 1942; and F. Campus and C. Massonnet in Brussels, Belgium, in 1956.

The American researcher F. R. Shanley, based on tests on aluminum–alloy columns carried out circa 1947, showed that the tangent modulus buckling stress theory, first proposed by Engesser in 1889, was a more correct solution for short columns than Engesser's reduced modulus theory proposed in 1895 (Gerard 1962; Shanley 1947; Bleich 1952, pp. 16–19). Shanley thought that Engesser's assumption, in the reduced modulus theory, that the column remains straight, was not correct. Shanley considered the simplified column as shown in Figure 2-6 [from Bleich

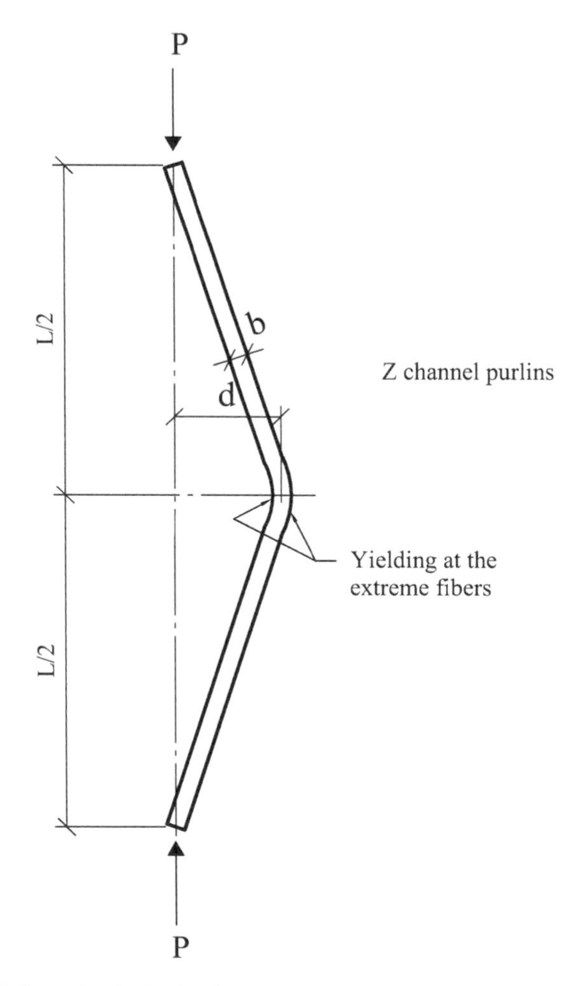

Figure 2-6. Column in the inelastic range.

Source: Reproduced from Bleich (1952, p. 17), courtesy of McGraw-Hill.

(1952, p. 17)] assuming an inelastic portion at midheight. Shanley derived the following relationship:

$$\rho = \rho_t \left(1 + \frac{1}{\left(\frac{b}{2d} + \frac{1+\tau}{1-\tau} \right)} \right) \tag{2-14}$$

where

ρ_t = Tangent modulus load,
b = Column width,
d = Deflection, and
τ = E_t/E, in which E_t is the tangent modulus and E is the Young's modulus.

The configuration, shown in Figure 2-6, is theoretically stable, similar to the deflected shape of an elastic column, when Euler's load is marginally exceeded. Shanley's work was endorsed by von Kármán, who was of the opinion that the tangent modulus load (first proposed by Engesser in 1889) should be used as the formula to determine the critical column load, because it was only slightly lower than the column critical load.

The generalized Euler or tangent modulus formula, proposed by Engesser, as previously mentioned, considered the stress–strain diagram that included yielding at the extreme fibers, using the tangent to the stress–strain curve as the tangent modulus, E_t. This is represented in Figure 2-7 (derived from Gerard 1962, p. 18).

When the slenderness ratio is such that the proportional limit is attained, the reduced stiffness for imperfect inelastic columns may be represented by substituting E for E_t. Thus, the generalized Euler or tangent modulus formula, applicable to perfect inelastic columns, is as follows:

$$P_{cr} = \frac{\pi^2 E_t I}{L^2} \tag{2-15}$$

This may be written in the following form (Bleich 1952, p. 15):

$$P_{cr} = \frac{\pi^2 E_t A}{(L/r)^2} \tag{2-16}$$

where r is the radius of gyration.

In 1944, the Column Research Council (CRC), which subsequently became the Structural Stability Research Council, was established in the United States to resolve differences in practice and opinions on stability problems. The first meeting took place on September 26, 1945, with 32 representatives from 17 organizations.

In 1948, the American Institute of Steel Construction (AISC), rather than using the tangent modulus formula or the secant formula (see Section 2.3),

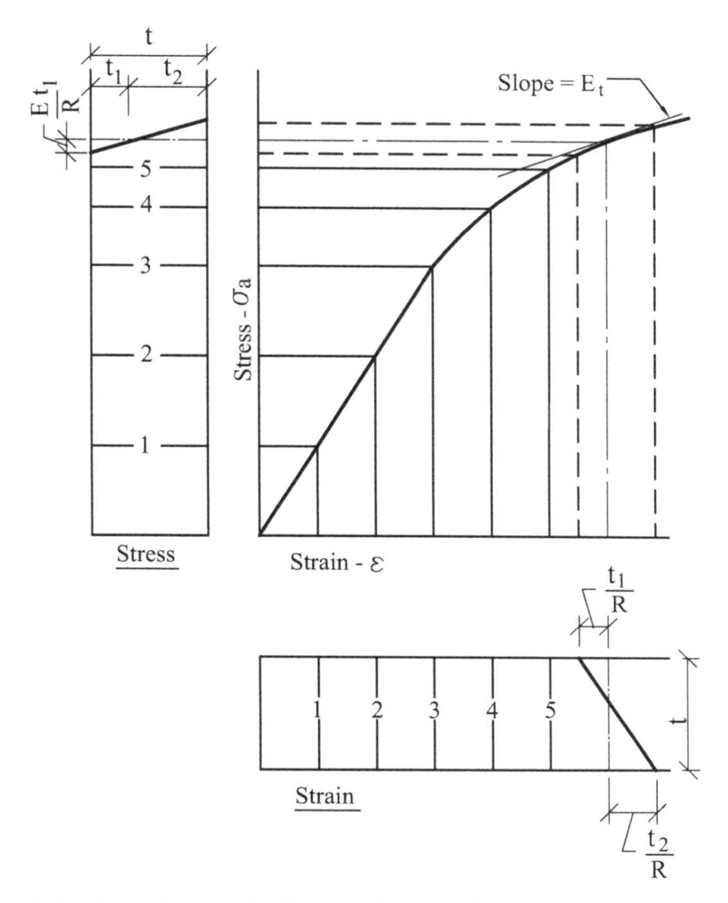

Figure 2-7. Strain and stress distributions for an inelastic column with small initial imperfections—tangent–modules model.

Source: Reproduced from Gerard (1962, p. 18), courtesy of McGraw-Hill.

approximated the curve, for short columns with slenderness ratios less than 120, using the parabolic equation as follows (Popov 1959, AISC 1948):

$$\frac{P}{A} = 17{,}000 - 0.485\left(\frac{L}{r}\right)^2 \quad \text{psi, a slenderness ratio, } L/r < 120 \tag{2-17}$$

For bracing and other secondary members with L/r between 120 and 200 (AISC 1948):

$$\frac{P}{A} = \frac{18{,}000}{(1 + L^2/18{,}000r^2)}\ \text{psi} \tag{2-18}$$

For main members, again for L/r between 120 and 200, a further reduction factor was applied as follows (AISC 1948):

$$\frac{P}{A}=\frac{18{,}000}{(1+L^2/18{,}000r^2)}\left(1.6-\frac{(L/r)}{200}\right)\text{psi} \tag{2-19}$$

Research on the effects of residual stresses because of hot rolling and welding, reducing the column strength, was well researched commencing in the late 1940s under the guidance of the CRC (Johnston 1976, p. 53). Residual stresses can vary significantly across the section and can be as much as 20 ksi. In the case of wide flange members, the tips of the flanges and much of the web are in compression and the remaining center portions of the flanges are in tension. Figure 2-8 indicates the residual stress distribution derived from Johnston (1976, p. 55) for a heavy wide flange member (W14 × 426). Also, refer to Chapter 4, Section 4.1, for more discussion on residual stresses arising from steel production. The effects of residual stress more prominently occur in shorter columns and less so for slender columns where performance is governed by Euler buckling. CRC included the effects of residual stresses in the design formula in 1960. From research work carried out by G. A. Alpsten, L. Tall, and J. Brozzetti in 1970, it was found that significant residual stresses, because of the welding of sections, could appreciably affect column strength. The residual stresses, because of welding, modify the residual stresses in the rolled plates. Stresses as high as yield could occur because of welding. Columns fabricated by welding with higher-strength steels are less affected than those made with lower-strength steels (Johnston 1976, pp. 57–58).

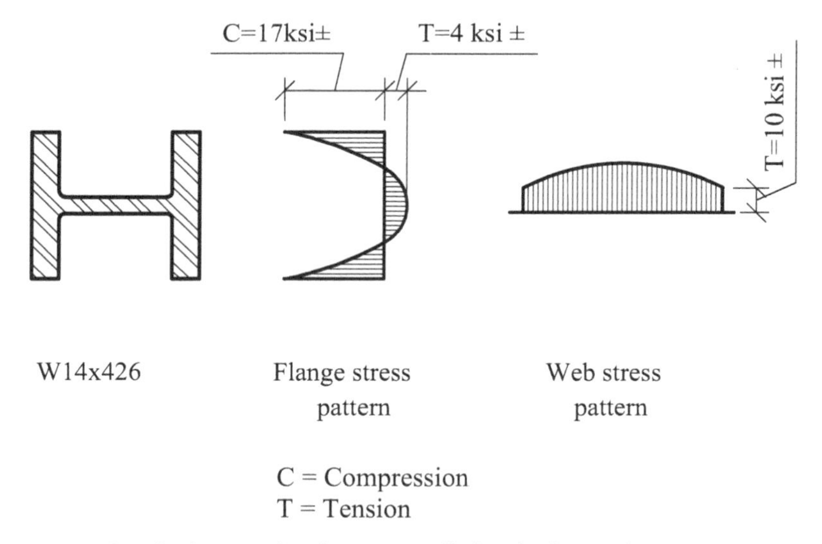

Figure 2-8. Residual-stress distribution in rolled wide-flange shapes.

Source: Reproduced from Johnston (1976, p. 55), courtesy of Wiley.

In Europe, the European Convention for Constructional Steelwork (ECCS) was established in 1959. A committee on stability was formed to help reduce the disparity on buckling in various national codes in Europe. Arbitrary safety factors had been applied to account for geometric imperfections and variable material properties that led to significant differences between the codes.

Extensive experimental and theoretical research was undertaken by ECCS that led to formulas based on a modified Perry formula. Although previous codes had a single strut curve, it was considered that several curves were required to address significant differences in performance between different steel sections, accounting for variations in initial deformations. Different sections include rolled sections, welded sections, and bolted sections. This led to the adoption of multiple column curves in some European Countries circa 1976. The curves were obtained from the results of tests on different column sections and statistical evaluation.

In the United States, it was considered that multiple curves would be overly complicated, particularly considering the different grades of steel, and, thus, this methodology was not adopted. For more discussion on residual stresses, see Section 3.1.

After much deliberation by the CRC in the United States, which is well-discussed in Johnston (1976), the following equations for allowable stress were presented, and these are given in the AISC's seventh edition, *Specification for the Design, Fabrication and Erection of Structural Steel for Buildings* dated February 12, 1969:

Inelastic

$$\frac{P}{A} = F_{a=} \cdot \frac{\left| 1 \cdot \frac{(KL/r)^2}{2C_c^2} \right|}{\frac{5}{3} + \frac{3(KL/r)}{8C_c} - \frac{(KL/r)^3}{8C_c^3}} \tag{2-20}$$

where $C_c = \sqrt{2\pi^2 E/F_y}$ and K is the effective length factor where $KL/r \leq C_c$

Elastic

where $KL/r > C_c$

$$\frac{P}{A} = F_a = \frac{12P^2E}{23(KL/r)^2}$$

which is based on Euler's formula with a factor of safety (FS) applied:

For the inelastic formula, FS is the denominator such that the FS varies from 1.67 for $KL/r = 0$ to 1.92 (or 23/12) for the elastic formula when $KL/r = C_c$.

Load resistance factor design (LRFD) was introduced by AISC in 1986 (AISC 2016c). The column curves were based on original research carried out by R. Bjorhovde (1988). Bjorhovde discusses the investigation of the factors affecting column strength for heavy columns. These included the restraint offered by the beam-to-column connection, the column length, the magnitude and distribution of residual stresses, and the initial out-of-straightness of the column. Consideration was given to the amount of out-of-straightness, and it was decided, based on steel-making practice, that L/1500 was appropriate to establish design curves. Three column curves were originally proposed by Bjorhovde to represent HSS members, wide flange members, and welded builtup wide flange members. Subsequent considerations by the AISC again led to only one curve's being adopted. These primarily pertained to the effecting of changes in industry practices, including welded builtup wide flange members, being no longer fabricated and improvements in the steel-making practice, resulting in better defined properties that helped reduce variations in steel properties. Minor modifications to address changes to the resistance factor from 0.85 to 0.90 were also made. The LRFD column formulas have not been reproduced in this book, as they are provided in ANSI/AISC 360 (AISC 2016c), which is readily available.

2.3 ECCENTRICALLY LOADED COLUMNS

A relatively simplified method that involved considering an eccentrically loaded column, similar to that shown in Figure 2-9, and developing a formula known as

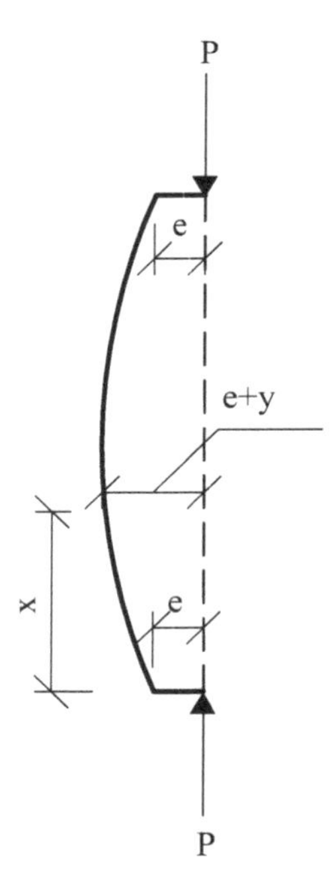

Figure 2-9. Eccentrically loaded column.
Source: Derived from Bleich (1952, p. 28).

the *secant formula* was proposed in 1858 (Johnston 1976). It took into account the initial out-of-straightness of the columns.

With reference to Bleich (1952, p. 45), the formula is based on the assumption that the deflected form, for a load P applied at an eccentricity of e, is as follows:

$$y = \left(\frac{\cos^{\alpha}((l/2) - x)}{\cos(\alpha L/2)} - 1 \right) e \tag{2-21}$$

where $\alpha = \sqrt{P/EI}$, and x and y are shown in Figure 2-9.

This leads to what became the well-known secant formula for columns, which is

$$\sigma = \frac{P}{A}\left(1 + \frac{ec}{r^2} \sec \frac{L}{2r} \sqrt{\frac{P}{EA}} \right) \tag{2-22}$$

where σ is the maximum compressive stress, and c is the distance of the extreme fibers from the centroidal axis, A = area.

For short columns, this reduces to

$$\sigma_{\max} = \frac{P}{A} + \frac{Mc}{I}$$

The secant formula for columns, considered to be conservative, in general, is referenced in Bleich (1952, pp. 45–47), Johnston (1976, p. 40), Popov (1959, p. 353), and Timoshenko and Gere (1961). Bleich (1952) was of the opinion that, because the formula was based on an arbitrary criterion, there could be a wide deviation between actual critical loads and calculated critical loads based on the secant formula.

Eccentric loads causing moments in the same plane of buckling were likely first investigated by the German engineer, A. Ostenfeld, in 1898 (Bleich 1952, p. 25). Further work by von Kármán was performed when he carried out a rigorous analysis that indicated the significant effects of slight eccentricities on the critical loads for short and medium-length columns. Curves were established that indicated the sensitivity of eccentricity. Figure 2-10, derived from Timoshenko and Gere (1961, p. 173), indicates the sensitivity for a column with $L/r = 75$. From Figure 2-10, it can be seen that a load increase is required to produce an increase in deflection. After a certain limit, where the curves descend, further deflection occurs as the load decreases. These limits indicate the maximum load that the column can sustain for a given eccentricity.

H. M. Westergard and W. R. Osgood, in their analytical presentation, given in 1928 on eccentrically loaded columns, used a modified version of von Kármán's method, adopting a part of a cosine curve. E. Chwalla, also using von Kármán's concepts, in a series of papers between 1928 and 1937, made accurate calculations based on the deflected form indicated in Figure 2-9, derived from Bleich (1952, p. 28). Chwalla produced tables and diagrams for eccentrically loaded columns, taking into account the column cross section, slenderness ratio, and eccentricity

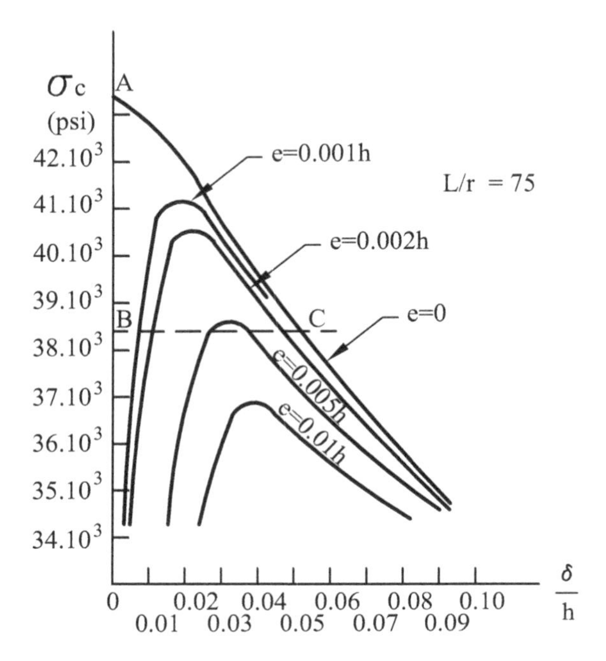

σ_c = Direct compressive stress
δ = Deflection
h = Depth of member

Figure 2-10. Eccentrically loaded column L/r = 75.
Source: Reproduced from Timoshenko and Gere (1961, p. 173), courtesy of McGraw-Hill.

(see Figure 2-11, derived from Bleich 1952, p. 31). K. Jezek, in 1934, proposed an exact and approximate method for different materials. Jezek considered the stress–strain curves with yield points of the materials and assumed, for simplification, a half sine curve for the deflected form from which he produced diagrams, tables, and design formulas (Bleich 1952, pp. 26–35). Ros and Brunner carried out tests, on 18 I beams subjected to eccentric loads, circa 1936. The results showed some correlation with the formulas developed. Although Bleich preferred the methods developed by Jezak, the secant formula, previously discussed, became the predominant design method for eccentrically loaded columns.

2.4 BRACING OF COLUMNS

A. F. S. Jasinsky, a Polish professor and engineer, took a deep interest in the buckling of columns and how to address the stability of lattice-type bridges

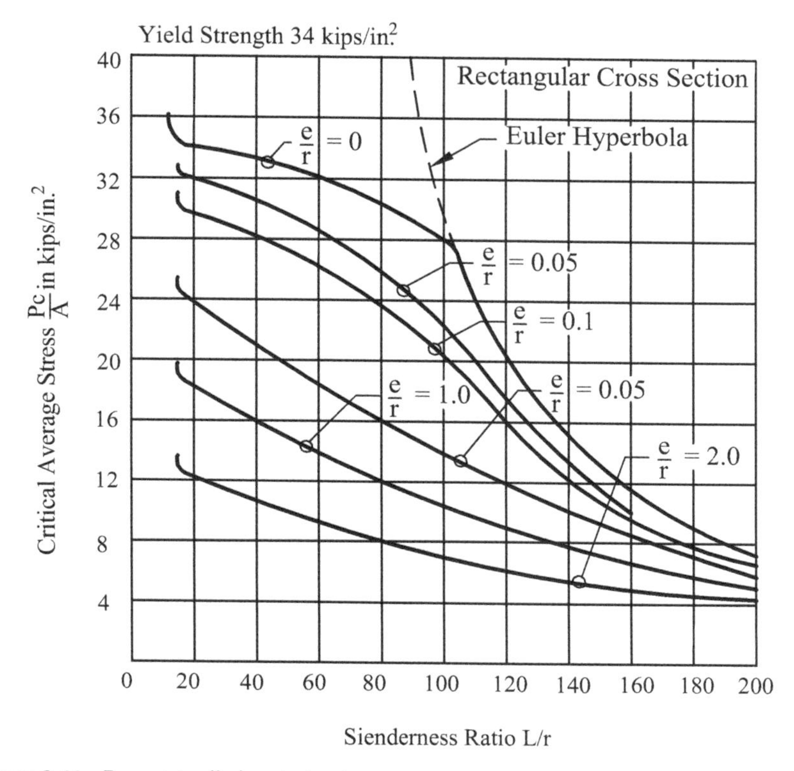

Figure 2-11. Eccentrically loaded column.

Source: Reproduced from Bleich (1952, p. 31), courtesy of United Engineering Foundation.

depicted in Figure 1-7. Before his untimely death, Jasinsky was able to establish rational procedures, based on a rigorous solution, to address lateral buckling of the top chord of these bridges, subjected to compressive forces.

As demonstrated by G. Winter, both adequate strength and stiffness are necessary to provide for lateral bracing. Essentially, strength is required to address initial out-of-straightness, and stiffness is needed to maintain stability (Winter 1960).

The need for adequate stiffness for bracing of a column is shown in Figure 2-12 (derived from Timoshenko and Gere 1961, p. 83). Considering the equilibrium of forces, the ideal condition for the bracing springs constant is as follows:

$$P_{cr} = \beta L \qquad (2\text{-}23)$$

where β is the stiffness constant ($K/''$, kN/mm).

For an intermediate brace with reference to Winter (1960),

$$P_{cr} = \beta L/2 \qquad (2\text{-}24a)$$

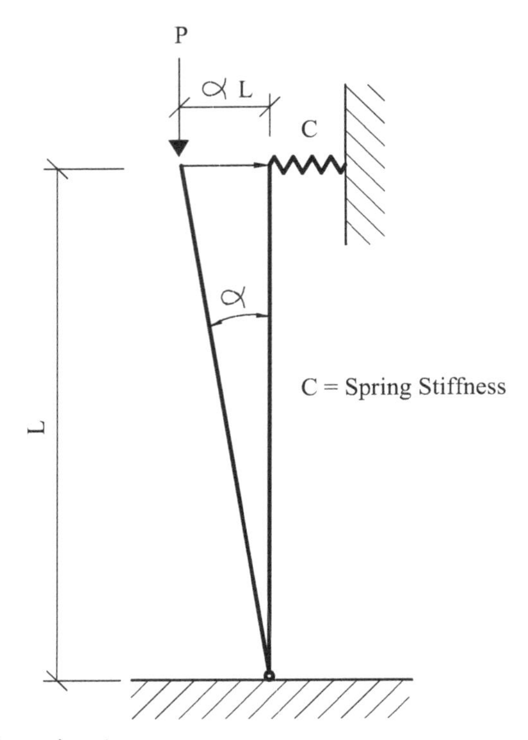

Figure 2-12. Column bracing.
Source: Reproduced from Timoshenko and Gere (1961, p. 83), courtesy of McGraw-Hill.

For many braces, with reference to Yura (1999),

$$P_{cr} = \beta L/4 \tag{2-24b}$$

Pincus (1964) considered the lateral support requirements for inelastic columns. With reference to the model shown in Figure 2-13 (from Pincus 1964), the following relationship was obtained:

$$P = \frac{4\beta}{L} + \frac{kL}{4} \tag{2-25}$$

where β is the rotational spring stiffness, and k is the lateral spring stiffness.

Pincus showed that, if the column is behaving inelastically, it requires a greater lateral stiffness than when behaving elastically.

An excellent discussion on bracing is given by Yura (1993) and Yura and Helwig (2001). Considering an initial out-of-straightness Δ_o at the midheight of the column, a relationship can be established between the axial load and the total lateral displacement Δ_T for the ideal brace, as shown in Figure 2-14(a, b),

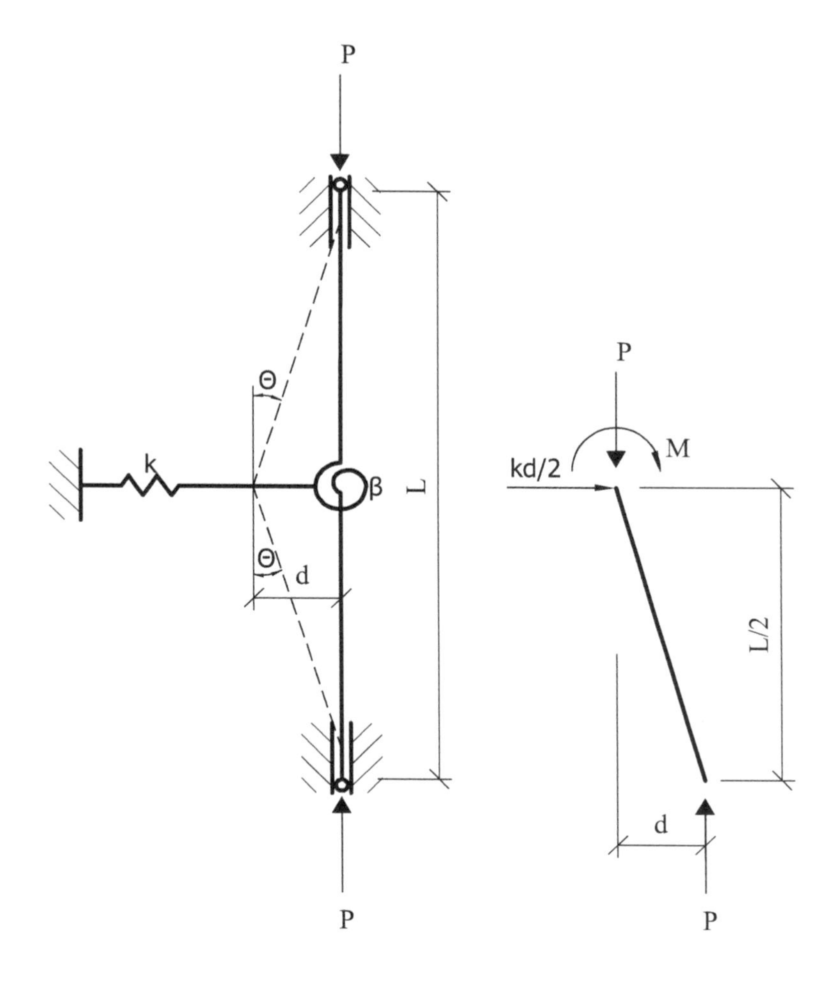

k = Lateral spring stiffness

Figure 2-13. Model representing lateral supports for inelastic columns.
Source: Reproduced from Pincus (1964), courtesy of AISC.

derived from Yura (1993). It can be seen that when a load is applied within 5% of P_e, $\cdot \Delta_T = 20\Delta_o$. As the load approaches the critical load, the lateral displacement, Δ_t, becomes very large. By increasing the stiffness by a factor of 2, $\Delta_T = 2\Delta_o$, and for $P = P_e$, the lateral displacement can be controlled. Thus, Yura demonstrated the importance of stiffness, pointing out that ideal stiffness was not sufficient and there was a need to overdesign the stiffness of the lateral bracing.

Yura and Helwig (2001), along with their excellent and highly informative discussion on bracing for stability, provide many examples on addressing stability issues.

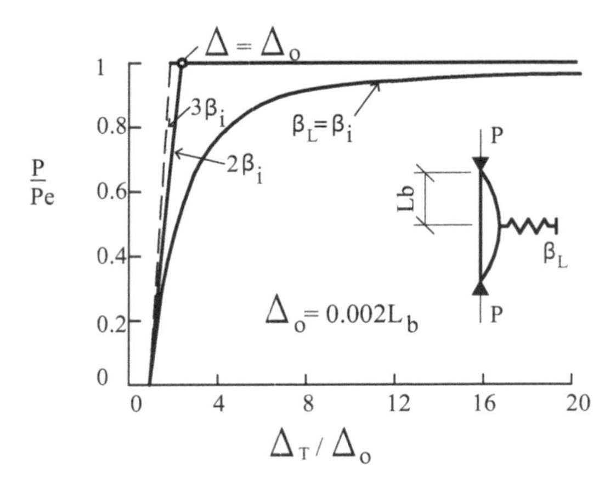

Figure 2-14.(a) Comparison of load/deflection for different brace stiffnesses.

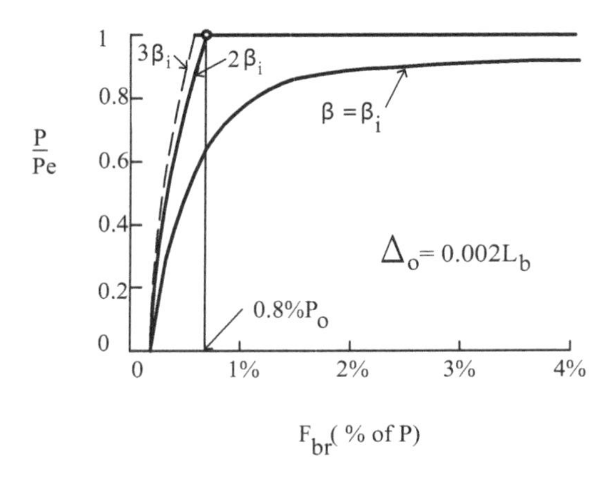

Δ_T = Total displacement at mid height
F_{br} = Force in brace
B = 2Pe / L (ideal brace stiffness)
$Pe = \pi^2 EI / Lb^2$

Figure 2-14.(b) Comparison of load/brace force for different brace stiffnesses.
Source: Reproduced from Yura (1993), courtesy of AISC.

2.5 INSTABILITY OF BEAMS/GIRDERS

A. Duleau conducted extensive experiments in France in the early nineteenth century on the strength of builtup beams within the elastic range. Much progress had been made in the understanding of the strength of materials since the eighteenth century, particularly in France, with C. L Navier's book documenting the progress made in understanding the strength of materials published in 1826.

Duleau's tests helped to gain a better understanding of evaluating the influence of factors such as beam, depth, shear, the connection of components with regard to the rigidity of beams, and so on. Although these did not have any indicators regarding beam stability, included in these experiments were tests on torsion using prismatic iron bars and circular shafts. Duleau concluded that tubular sections performed well compared with prismatic sections and also found that, for circular sections, the torsional shear stresses were proportional to the distance from the neutral axis. This was not so for prismatic bars (Timoshenko 1953).

Subsequent improvements to Duleau's work on torsion were made by the mathematician, engineer, and physicist A. L Cauchy. Theories on torsion were established in the mid-nineteenth century by the French researcher, Barré de Saint-Venant, with his proposal to adopt the procedure known as the *semi-inverse method* (Timoshenko 1953). As found earlier by Duleau, pure torsion occurs only with circular shafts, with cross sections remaining in plane, which can be readily calculated based on the torsional rigidity (GJ, where G is the shear modulus and J is the Saint-Vernant torsion constant or torsional inertia). In the case of circular shafts, distortion along the longitudinal axis of the member does not occur. Saint-Venant recognized that warping of the cross section occurs, as shown in Figure 2-15. In the case of the I beam shown in the figure, no significant shearing stresses occur along the y-axis (web of the member), with bending occurring in the flanges in the horizontal plane. Thus, it became important to understand the two components to torsional resistance, that is, pure torsion and warping (Timoshenko 1953).

According to Johnston (1976, p. 121), Fairbairn carried out beam tests in 1854, resulting in the conclusion that compression flanges that are wider and

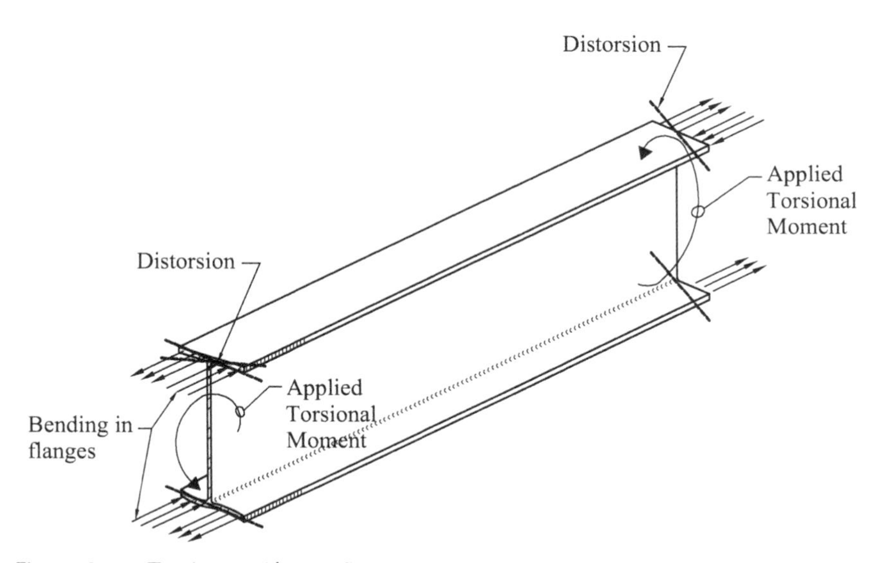

Figure 2-15. Torsion on I beam (isometric view).

thicker than tension flanges had improved their buckling strength. Subsequently, tests on steel beams by Burr in 1884, Marburg in 1909, and Moore in 1910 helped to develop design formulas (Johnston 1976, p. 121).

A. G. Greenhill, circa 1881, using functions determined by the famous astronomer Bessel, solved a variety of buckling problems, including the buckling of bars under distributed axial loads, as shown in Figure 2-16 (Timoshenko and Gere 1961, p. 101). This is similar to the varying compressive forces that occur in beam flanges.

The differential equation derived is as follows:

$$EI\frac{d^2y}{dx^2} = -q(L-x)\frac{dy}{dx} \tag{2-26}$$

where q is the uniformly distributed load for the condition shown in Figure 2-16.

Solving the aforementioned equation yields the following approximate equation (Timoshenko and Gere 1961, p. 105):

$$(qL)_{cr} = \frac{7.837EI}{L^2} \tag{2-27}$$

where $(qL)_{cr}$ is the critical uniformly distributed load for the condition shown in Figure 2-16.

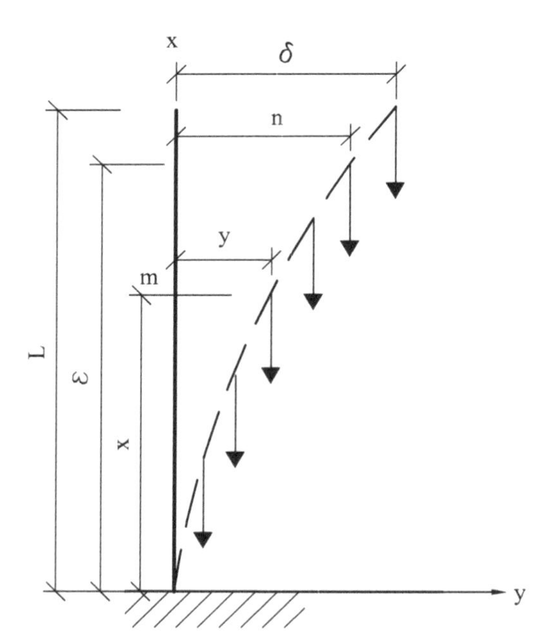

Figure 2-16. Buckling of a column under its own weight.

Source: Reproduced from Figure 2.39 in Timoshenko and Gere (1961, p. 101), courtesy of McGraw Hill.

Ludwig Prandtl carried out tests at the Munich Polytechnic Institute around 1899 on beams with a narrow rectangular cross section bent around their major axis (Johnston 1976, p. 121). Prandtl was able to develop a theory for narrow rectangular beams that accurately determined the critical load at which lateral buckling occurred in the absence of lateral support at the compression flange. Prandtl established the following (Timoshenko and Gere 1961, p. 257):

$$(Mo)_{cr} = \frac{\pi}{L}\sqrt{EI_n C} \tag{2-28}$$

where $(Mo)_{cr}$ is the critical moment

$$C = GJ = 1/3hb^3G$$

where
h and b = Dimensions of the rectangular section,
G = Shear modulus, and
J = Torsion constant.

The same solution was derived by A. G. M. Michell, also in 1899, who also carried out tests on steel bars that were consistent with the theory put forward by Prandtl (Bleich 1952, pp. 149–152).

F. S. Jasinsky solved problems with the lateral buckling of compression diagonals of lattice trusses. Jasinsky considered the lateral stiffness of the compression member of the truss by other members of the truss, as shown in Figure 2-17.

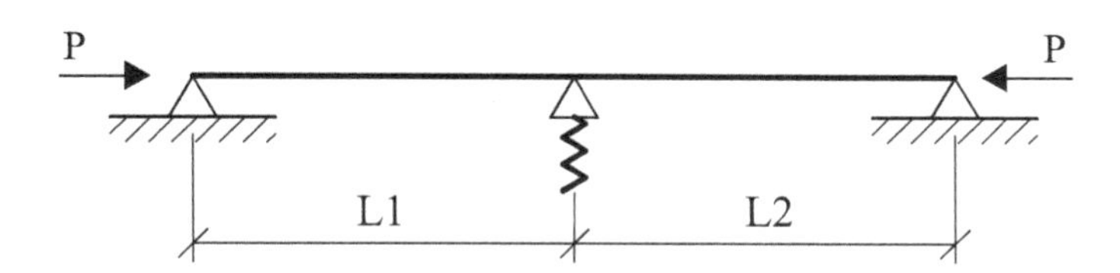

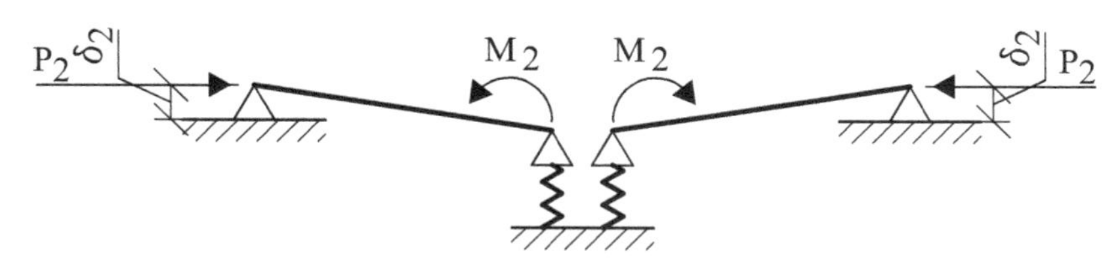

Figure 2-17. Continuous strut.

Source: Derived from Timoshenko and Gere (1961, p. 70).

The critical load, obtained for this condition (Timoshenko and Gere 1961, p. 72), is as follows:

$$P_{cr} = \frac{\pi^2 EI}{(L_1 + L_2)^2} \tag{2-29}$$

The great professor, Stephen P. Timoshenko, at the Polytechnique Institute in Kiev circa 1910, began working on the lateral buckling of I-beams (Timoshenko 1953). Timoshenko, who is credited with being the first to provide solutions for I-beams, derived the fundamental differential equation of the torsion of symmetric I-beams and investigated the lateral buckling of transversely loaded deep I-beams. Timoshenko solved the problem of stability in 1913 (Bleich 1952, p. 150). Essentially, as loading of an I-beam without lateral support is applied, causing bending around the major axis, the beam remains stable provided it is below a critical value. When the load is increased, the beam slightly twists but can remain stable (Figure 2-18). As explained in the classic book *Theory of Elastic Stability* by Timoshenko and Gere (1961), on the load attaining the critical load, the beam becomes unstable as the twisting, caused by compression in one of the flanges, cannot be controlled. Timoshenko and Gere (p. 254) established, for thin-walled sections, the following differential equation:

$$C_1 \frac{d^4\phi}{dz^4} - \frac{Cd^2\phi}{dz^2} - \frac{M_o{}^2\phi}{EI_n} = 0 \tag{2-30}$$

where

$C = GJ$ is the torsional rigidity in which G is the Shearing modulus of elasticity and J is the torsion constant,
$C_1 = EC_w$ is the warping rigidity,
$M_o =$ Applied moment in the vertical plane, and
$I_n =$ Moment of inertia around the minor axis

Timoshenko and Gere (p. 267), based on a rigorous analysis for uniform loading and taking into account both pure torsion and warping, evaluated the critical load, qL, as follows:

Simply supported I-beam

$$(qL)_{cr} = \gamma_4 \frac{\sqrt{EI_n C}}{L^2} \tag{2-31}$$

where γ_4 depends on the ratio L^2C/C_1 and the position of the load.

This assumes that the loads are applied at the shear center and that adequate lateral restraint is applied at the ends of the beam, which accounts for restraint to buckling being provided by less critical elements. A variation of the moment affects this, as discussed in Section 2.6.

F. Bleich, in 1935, further extended the stability condition in a general form to apply to unequal flanges and also addressed the condition in which the tension flange is restrained from lateral displacement (Bleich 1952, p. 150). Tests

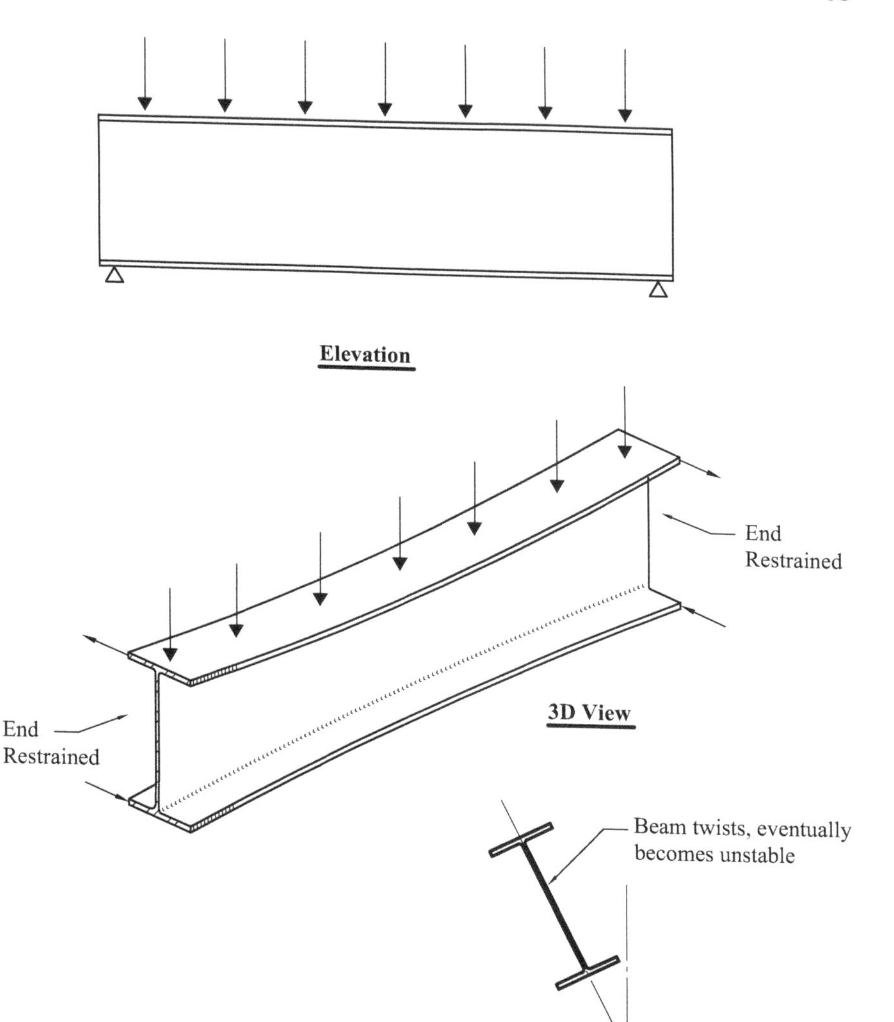

Figure 2-18. Instability of beams/girders.

on I-beams were carried out by H. F. Moore in 1913 and later in 1932 on light beams by M. S. Ketchum and J. D. Draffin. The tests indicated much higher critical stresses than those derived from theory.

Tests by C. Dumont and H. N. Hill on rectangular aluminum beams, carried out in 1937, showed good agreement with theory (Bleich 1952, p. 152).

G. Winter established, in 1941, approximate formulas for lateral buckling of nonsymmetrical I beams. A more exact solution assuming constant bending moment was given by H. N. Hill in 1942 (Bleich 1952). Further contributions were made by H. Nylander, J. N. Goodier, and K. De Vries in the 1940s.

For a discussion on the bracing of beams/girders, see Section 2.7.

2.6 VARIATION OF THE MOMENT ALONG THE BEAM

For the elastic condition, with a uniform moment applied to an I-beam, without axial load and assuming that the loads are applied at the shear center, Timoshenko and Gere (1961, p. 250) established the critical moment, M_{cr}, to be as follows:

$$M_{cr} = \frac{\pi}{L}\left[(EI_Y GJ)\left(1 + \frac{C_1 \pi^2}{GJL^2}\right)\right]^{1/2} \tag{2-32}$$

where

E = Young's modulus,
G = Shear modulus,
M_a = Unbraced length,
M_b = Weak axis moment of inertia, and
M_c = Warping constant.

With reference to Timoshenko and Gere (1961) and Bleich (1952), formulas were also established for inelastic buckling of I-beams, by also considering the loading applied at the top flange as well as at the centroid.

With reference to Wong and Driver (2010), which includes an overview of the literature on this subject, methods were developed for unequal end moments by Salvadori (1955) and Austin (1961) as follows:

$$C_b = 1.75 + 1.05k + 0.3k^2 \leq 2.3 \quad [\mathit{Salvadori}(1955)] \tag{2-33}$$

$$C_b = (0.6 - 0.4k)^{-1} \leq 2.5 \quad [\mathit{Austin}\ (1961)] \tag{2-34}$$

where C_b represents the equivalent uniform moment factor, and k quantifies the influence of the flange force variation between the two ends.

Methods for special moment distributions were also developed by Clark and Hill (1960), based on the minimum potential energy. Nethercot and Trahair (1976) developed procedures and later equations, for a general moment distribution, were developed by Kirby and Nethercot (1979).

Different equations have been used in different codes and they vary significantly, according to Wong and Driver (2010). They found the Canadian Code to be grossly inaccurate and the equation given by AISC to produce nonconservative results in some cases. However, they found the British Code, BS 5950-1, to be conservative but not accurate. Wong and Driver proposed the following equations:

$$C_b = \frac{4M_{max}}{\sqrt{M_{max}^2 + 4M_a^2 + 7M_b^2 + 4M_c^2}} < 2.5 \qquad (2\text{-}35)$$

where M_a, M_b, and M_c are the moments at the quarter points

where

C_b = Lateral–torsional buckling modification factor for nonuniform moment diagrams when both ends of the unsupported segment are braced,
M_{max} = Absolute value of the maximum moment in the unbraced segment,
M_a = Absolute value of the moment at the quarter point of the unbraced segment,
M_b = Absolute value of the moment at the centerline of the unbraced segment, and
M_c = Absolute value of the moment at the three-quarter point of the unbraced segment.

The aforementioned equation is similar to the work presented by Wilkerson (2005) recommended for general use with distributed loads. Wilkerson (2005) also found that the 2005 AISC specification, in some cases, where the top flange was loaded to be nonconservative.

Wong and Driver's equations were accepted by the Canadian Standards Association.

2.7 BRACING OF BEAMS/GIRDERS

As stated in Section 2.4, both adequate strength and stiffness are required to provide for lateral bracing of the compression flange of the beam or girder.

Bracing of the compression flanges of beams and girders may initially be looked at simply as a column. However, whereas column buckling is a function of flexure with beam buckling, both flexure and torsion occur, affecting bracing requirements.

According to J. A. Yura (1993), bracing may be divided into two categories as follows: lateral bracing and torsional bracing.

2.7.1 Lateral Bracing

Lateral bracing prevents lateral displacement because of compression occurring from flexure.

With regard to I-shaped and channel sections, the most effective location for lateral bracing to prevent twisting is the compression flange. No vertical stiffeners are required when this occurs and adequate brace stiffness is provided.

Bracing, located at the midheight of the beam, is less effective and is totally ineffective if lowered to the bottom flange. An example of lateral bracing is shown in Figure 2-19, derived from Yura (1993). According to Yura (1993), if a brace is

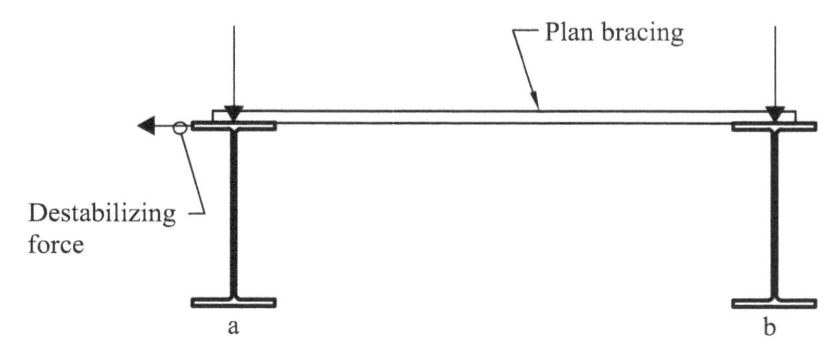

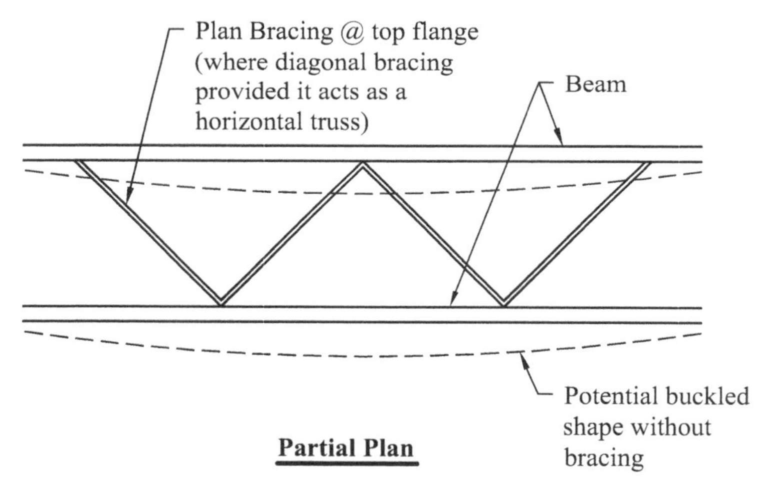

Figure 2-19. Lateral bracing.

Source: Reproduced from Yura (1993), courtesy of AISC.

lowered to the centroid of the beam with a stiffener plate, a significantly greater brace stiffness is required than for a brace located at the compression flange. The brace stiffness required further increases if a stiffener is not provided. These characteristics are shown in Figure 2-20, derived from Yura (1993).

2.7.2 Torsional Bracing

Torsional bracing, by directly providing adequate stiffness in the bracing system, prevents a twist of the cross section. An example of this is the case of upstanding

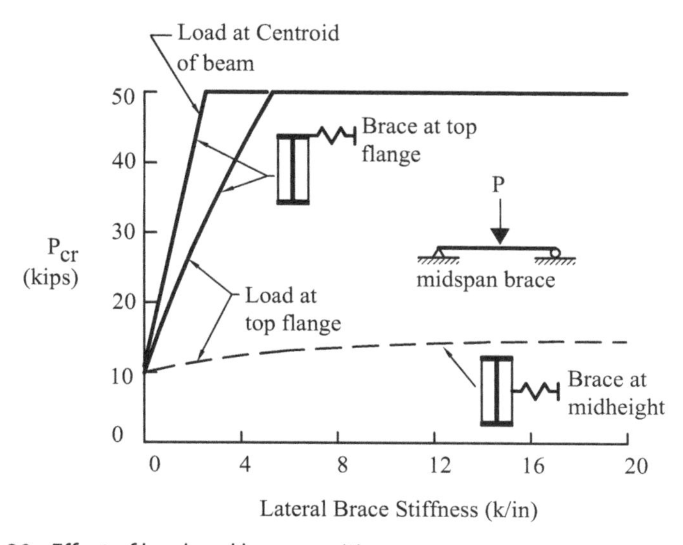

Figure 2-20. Effect of load and brace position.

Source: Reproduced from Yura (1993), courtesy of AISC.

girders shown in Figure 2-21(a, b), derived from Yura (1993), where the floor beam and vertical web stiffeners provide stiffening elements or the transverse bracing system to prevent the twist of the beam.

2.7.3 Effects of Load Application

Analytical studies have shown that if the load is applied directly at the top (compression) flange, greater instability occurs, which tends to reduce the effectiveness of the top flange brace [Figure 2-22(a)]. If the brace is lowered to the centroid of the beam, the brace is virtually ineffective with load applied to the top

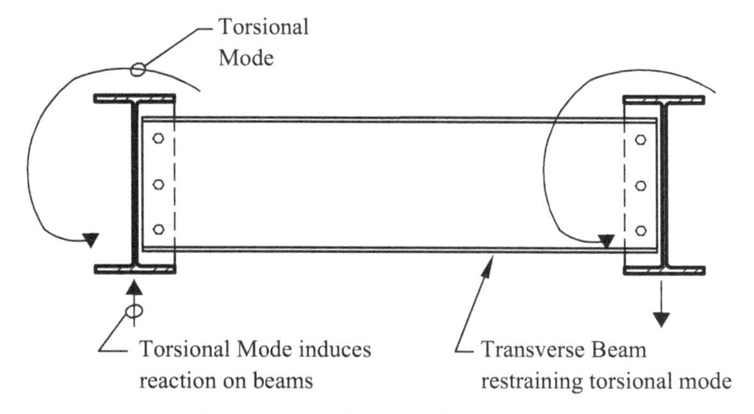

Figure 2-21.(a) Torsional bracing: Section showing transverse beam.

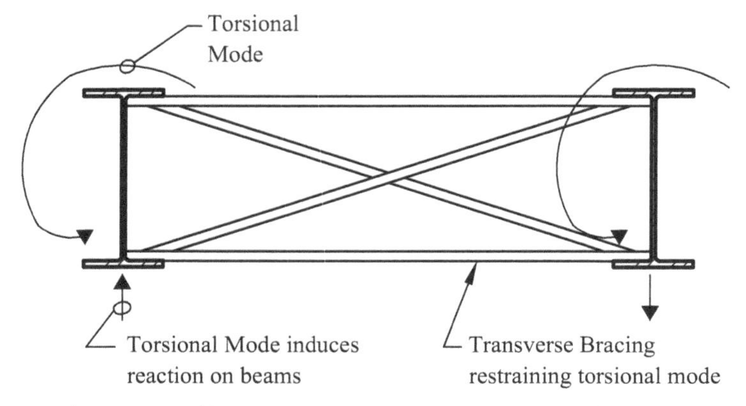

Figure 2-21.(b) Torsional bracing: Section showing transverse beam.
Source: Reproduced from Yura (1993), courtesy of AISC.

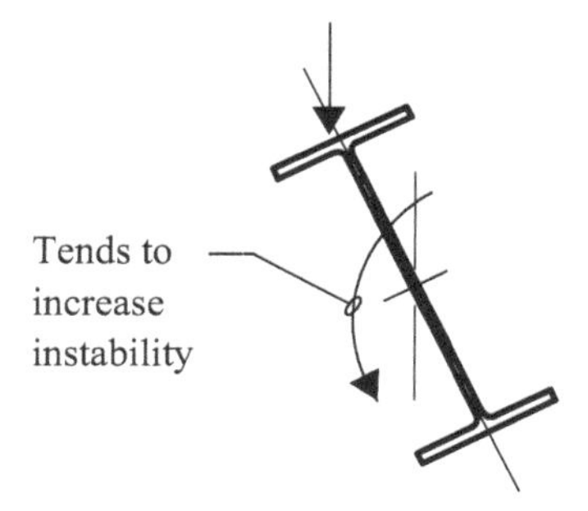

Figure 2-22.(a) Section loads applied at top.

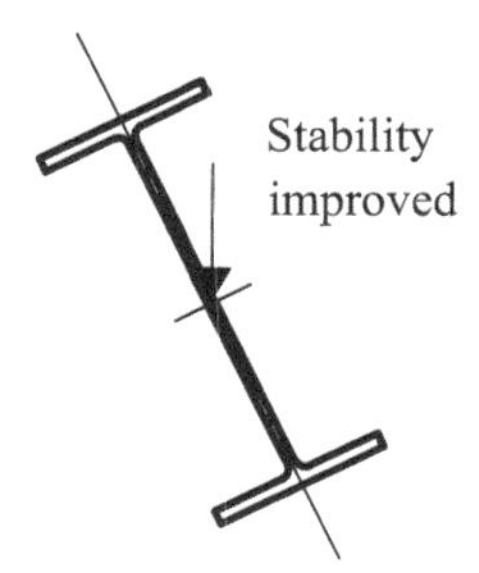

Figure 2-22.(b) Section loads applied at centroid.

flange [Figure 2-22(b)] unless stiffeners are provided. Applying the load at the centroid of the beam significantly improves the condition. If the load is applied directly to the bottom (tension) flange, this improves the condition even further because the application of the load tends to aid stability [Figure 2-22(c)].

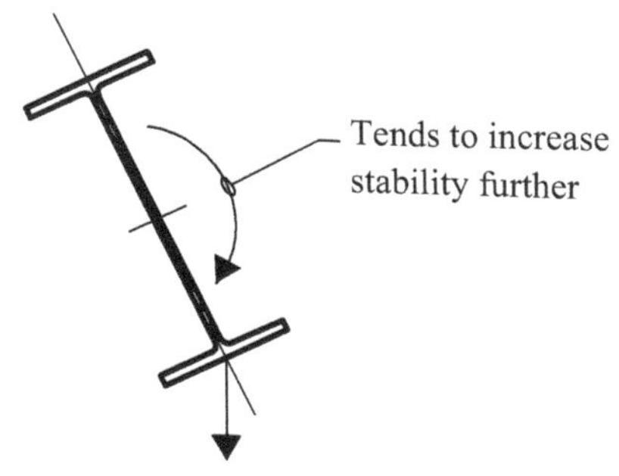

Figure 2-22.(c) Section loads applied at bottom.

2.7.4 Moment Gradient and Beams Subjected to Double Curvature

The basic case for the consideration of a beam is a uniform moment case with a constant compression in the compression (usually top) flange.

Where the moment is not uniform, such as with a simply supported beam subjected to uniform loading, the compression decreases from the center of the beam and, thus, results in fewer stability requirements. Procedures to represent the variation of the moment along the beam are discussed in Section 2.6.

Where beams are subjected to double curvature resulting in the compression flange changing from the top flange to the bottom flange, both flanges may need to be braced as bracing only one flange is ineffective. According to Yura (1993), beams with compression in both flanges have greater bracing requirements than a beam subjected to single curvature and braced at the top (compression) flange. Bracing stiffness requirements may be several times more for the double curvature case than for the single curvature case. Yura (1993) also emphasizes that, as has been considered by some engineers, it is incorrect to assume that the inflection point (zero moment) is a brace point.

2.7.5 Lateral Restraint

Lateral restraint, in the form of bracing or other means, needs to provide for both strength and stiffness. The introduction of a single brace, with sufficient strength and stiffness, to the top compression flange of the beam will change the buckling behavior from a single sine wave to a double sine wave. Likewise, having two braces will change the buckling behavior to a triple sine wave.

With reference to Yura (1993) [similarly stated in Yura and Helwig (2001)], brace strength $F_{br} = 0.008\rho$, where F_{br} is the brace strength required and ρ is the equivalent compressive beam flange force. Yura, based on Winter (1960), determined the stiffness requirement to be as follows:

$$\begin{aligned} &\beta_L = 2\#(C_b P_f) C_L C_d / L_b \\ \text{or} \quad &\beta_L = 2\#(M_f / h) C_L C_d / L_b \end{aligned} \tag{2-36}$$

where # = 4 – 2/n and n is the number of braces

$$C_b P_f = C_b \pi^2 E I_{ye} / L_b^2 \text{ or } M_f / h$$

where M_f is the maximum moment

$$C_L = 1 + 1.2/n \text{ for top flange loading}$$
$$= 1.0$$

$$C_d = 1 + (M_S/M_L)^2 \text{ for double curvature}$$
$$= 1.0 \text{ for single curvature}$$

The aforementioned is based on the beam/girder behaving elastically. Bracing requirements for both strength and stiffness are provided in the construction specification ANSI/AISC 360 (AISC 2016c).

2.7.6 Summary of Beam/Girder Bracing

In summarizing the aforementioned, it is important to acknowledge that beam/girder instability and bracing requirements are affected by brace location, brace stiffness, the number of braces, moment gradient, load location, and the properties of the girder.

Plastic design procedures in earlier AISC Specifications required "Members to be adequately braced to resist lateral torsional displacements at the plastic hinge locations associated with the failure mechanism" (AISC 1978). The commentary in AISC (1978) discusses the need to provide more bracing in a continuous frame subjected to plastic hinging than a continuous frame designed in accordance with elastic theory. AISC (1978)'s commentary advises that "when bending takes place about the strong axis, any W-shape member tends to buckle out of plane. It is for this reason that lateral bracing is needed."

Plastic design developed in Cambridge, England, starting in the early 1950s by Baker, Horne, and Heyman (Horne 1963), also led to recommendations that the lateral restraint of the plastic hinges was particularly essential at the first hinges formed in the collapse mechanism, such as at the eaves of single-story portal frames.

AISC (1978) contains a section dedicated to plastic design. It states that the members are to be "adequately braced to resist lateral and torsional displacements at the plastic hinge locations." It also requires laterally unsupported distance, L_{cr}, "from such braced hinge locations to similarly braced adjacent points" not to exceed the following:

$$\frac{L_{cr}}{ry} = \frac{1{,}375}{Fy} + 25 \quad \text{when} + 1.0 > \frac{M}{M_p} > -0.5 \tag{2-37}$$

$$\frac{L_{cr}}{ry} = \frac{1{,}375}{Fy} \quad \text{when} - 0.5 \geq \frac{M}{M_p} > -1.0 \tag{2-38}$$

For $Fy = 50$ ksi, Equations (2-38) and (2-39) give the following:

$$\text{For Equation (2-38),} \quad \frac{L_{cr}}{ry} \leq 52.5 \tag{2-39}$$

$$\text{For Equation (2-39),} \quad \frac{L_{cr}}{ry} \leq 27.5 \tag{2-40}$$

The current steel requirements for compact sections ANSI/AISC 360 (AISC 2016c) require limiting the unbraced length for yielding and also allow increased unbraced length where the required moment, M_r, is less than ϕM_p.

ANSI/AISC 360 (AISC 2016c) also has requirements for checking the adequacy of brace stiffness based on the significant work by Yura and Helwig (2001).

The equations in ANSI/AISC 360 (AISC 2016c) are not repeated as they are readily accessible.

With regard to seismic design, the requirements in ANSI/AISC 341 Seismic (AISC 2005b) are as follows.

Both flanges of beams shall be laterally braced with a maximum slenderness ratio of

$$\frac{L_{cr}}{ry} \leq 0.086E/F_y \tag{2-41}$$

For $F_y = 50$ ksi

$$\frac{L_{cr}}{ry} \leq 50$$

Lateral bracing adjacent to plastic hinges has to provide a required strength of

$$P_u = 0.06\, M_u / h_o \tag{2-42}$$

ANSI/AISC 341 Seismic (AISC 2016b) has a similar equation, allowing for material over strength and also includes requirements for checking the adequacy of brace stiffness.

The equations in ANSI/AISC 341 (AISC 2016b) are not repeated as they are readily accessible.

Placement of lateral bracing nearest to the plastic hinge of moment frame beams is based on ANSI/AISC 358 (2005a, 2016a). In the case of the reduced beam connection (RBS), based on experimental and analytical studies, supplemental lateral brace at the RBS may be omitted as long as the normal lateral bracing is provided.

In the opinion of the author, this appears questionable based on the work by Baker, Horne, and Heyman (Horne 1963) in the early 1950s, regarding steel moment frame connection testing performance and a consideration of out-of-plane concurrent forces.

For intermediate moment frame

$$\frac{L_{cr}}{r_y} \leq 0.17\,E/F_y \tag{2-43}$$

$$F_y = 50\,ksi$$

$$\frac{L_{cr}}{r_y} \leq 98.6$$

Again, these requirements have been updated with similar equations allowing for material overstrength and including requirements for checking the adequacy of brace stiffness.

The equations in ANSI/AISC 341 (AISC 2016b) are not repeated as they are readily accessible.

Ordinary moment frames

Lateral bracing requirements are to comply with ANSI/AISC 360 (AISC 2016c).

Braces are required to satisfy requirements in ANSI/AISC 360 (AISC 2016c), which include requirements related to both strength and stiffness. The equations in ANSI/AISC (AISC 2016c) are not repeated as they are readily accessible.

2.8 GLOBAL INSTABILITY OF FRAMED STRUCTURES

Much of the previous discussion pertains to individual components. Moment frame systems utilizing rigid or semirigid connections have been adopted in steel structures for well more than a century.

As a result, there has been much research and development on the subject of global instability of framed structures. Müller-Breslau, in Germany, presented procedures to address the buckling of the top chord of trusses for bridges in 1908, and the subject of buckling of frameworks was considered circa 1909, by H. Zimmermann.

In 1918, Engesser carried out small-scale model tests of elastically supported columns. In 1919, analysis procedures for the stability of two-dimensional plane frameworks with rigid joints, primarily for aircrafts, were developed by F. Bleich. The application of a reduced modulus allowed the theory to be used in the elastic and plastic ranges of buckling. In 1919, establishing methods to determine the effective length of compression members in two dimensions was also attempted by F. Bleich. Later, in 1929, a general method for buckling problems related to space frameworks was published by F. Bleich and H. Bleich (Bleich 1952, p. 194).

In 1926, a similar approach to that by F. Bleich was developed by von Mises and J. Ratzersdorfer, again in Germany, by considering pin-connected trusses with parallel chords (Bleich 1952). In 1932, W. R. Osgood developed the stability condition for three-bay systems and, in 1934, applied Bleich's theory to unbraced compression members in aircraft frameworks (Bleich 1952, p. 194). In 1936, an approximate formula for the effective length of columns for frame structures was developed by M. G. Powein and also in 1937 by K. Borkman, who also produced charts. In 1937, a method for analyzing the stability of frameworks in the elastic range by considering translational and rotational restraints, was developed by W. Prager (Bleich 1952, p. 195). Using moment distribution, E. E. Lundquist, in 1937, developed procedures to determine the stiffness criteria for stability, known as the Lundquist series criterion.

E. Chwalla, in 1937, developed an accurate stability solution to a three-bay frame by considering an elastic–plastic stress–strain curve of structural steel (Bleich 1952, p. 225).

In 1941, E. Chwalla and F. Jokisen, utilizing the slope deflection method, studied the stability of multistory frames (Bleich 1952, p. 195). In 1951, Hoff carried out eight tests on truss models that validated failure loads, computed by Hoff's computational method previously mentioned (Bleich 1952, p. 196).

Adjustments for smaller or larger effective length had been known at least throughout the 1940s. According to Dumonteil (1999), in 1949 or before, it appears that approximate formulas were established by L. H. Donnell. N. M. Newmark also carried out work in 1944 that produced quite accurate K factor equations (Dumonteil 1999). An adaptation of Newmark's work by Jostein Hellesland resulted in the following formulas:

$$K = \sqrt{\frac{(G_A + 4/\pi^2)(G_B + 4/\pi^2)}{(G_A + 8/\pi^2)(G_B + 8/\pi^2)}} \tag{2-44}$$

where G_A and G_B are restraint factors [see Equation (2-49)] that produced quite accurate K factor equations (Dumonteil 1999). In the United States, the K factor was first introduced in the 1963 AISC specification (AISC 1963). It is important to understand the difference between multistory frames that are braced and those that are unbraced.

In the case of multistory frames that are braced, another lateral system (e.g., braced frames, concrete and/or masonry shear walls) must be present to provide lateral bracing for the building and the multistory frame(s) (Figure 2-23). For those that are braced in the same direction as the multistory frame, the effective length factor $K = 1.0$. In the case of those that are unbraced, it is important to understand that the building completely relies on these frames for stability as well as to provide resistance to lateral forces (e.g., wind or seismic) (Figure 2-24). As the building leans because of lateral forces, additional horizontal component forces, commonly known as "P-delta effects," occur that require resistance and sufficient stiffness from the multistory frames. For multistory frames, in the plane of the frame, $K \geq 1.0$.

Dumonteil (1999) discusses the historical background on the K factors. However, Dumonteil notes that the formulas, discussed in McGuire (1968), do

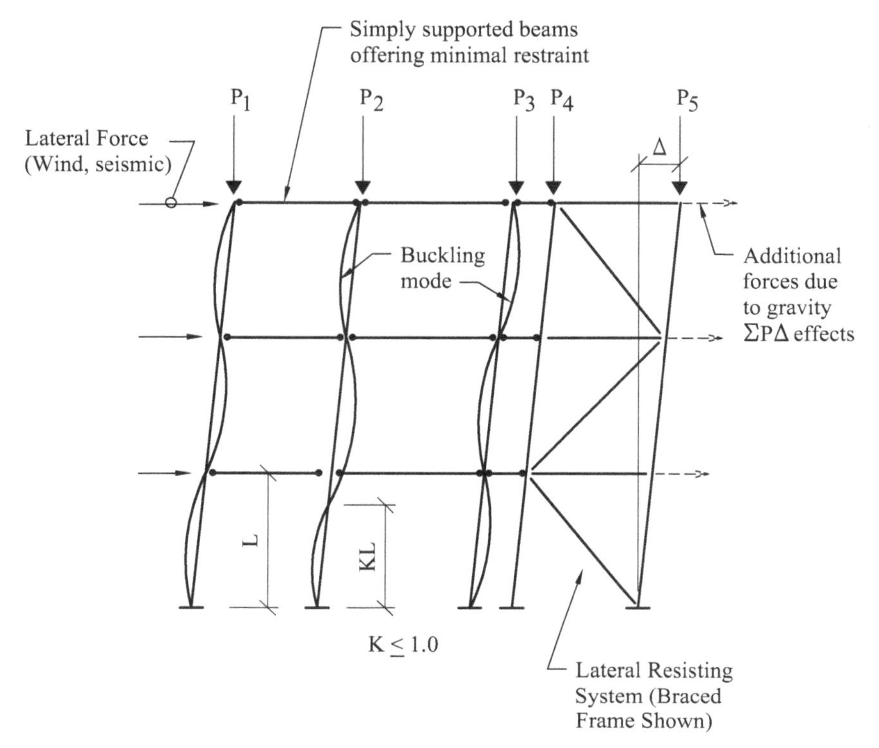

Figure 2-23. Multistory frame—braced.

not indicate the authorship for braced and unbraced frames. The formulas are as follows:

$$\frac{G_A G_B}{4}\left(\frac{\pi}{K}\right)^2+\left(\frac{G_A+G_B}{2}\right)\left(1-\frac{\pi/K}{\tan(\pi/K)}\right)+\frac{\tan(\pi/2K)}{\pi/2K}=1 \tag{2-45}$$

no sideway (braced frames)

$$\frac{G_A G_B(\pi/K)^2-36}{6(G_A+G_B)}=\frac{\pi/K}{\tan(\pi/K)} \tag{2-46}$$

for sway conditions (unbraced frames).

The aforementioned formulas make use of the restraint factors G_A and G_B at the two ends of the column section being considered. An end restraint factor G is defined as

$$G=\frac{\Sigma\left(I_c/L_c\right)}{\Sigma\left(I_b/L_b\right)} \tag{2-47}$$

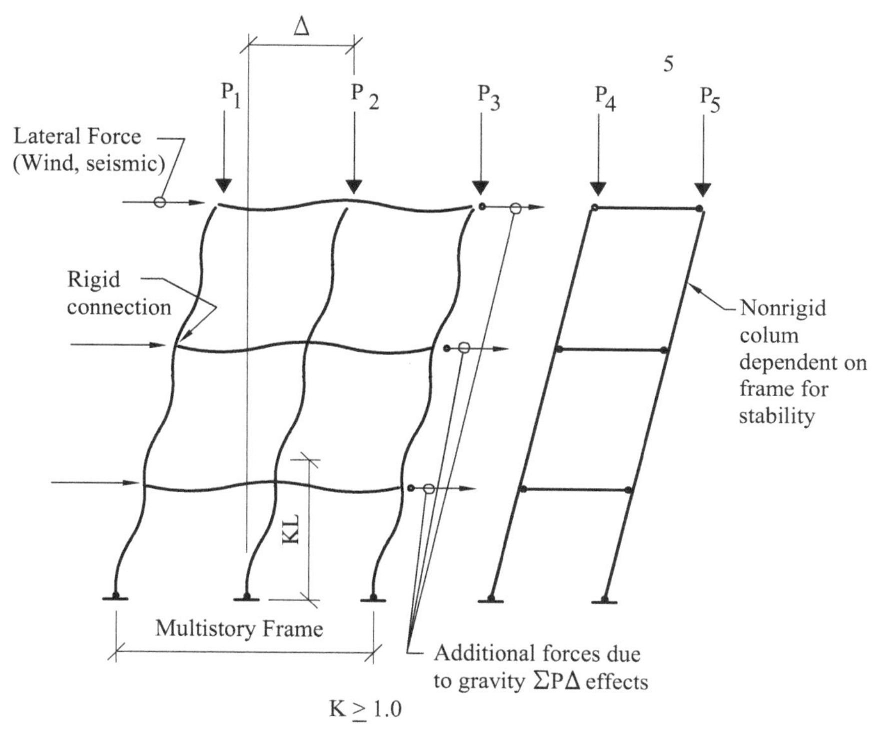

Figure 2-24. Multistory frame—unbraced.

Charts incorporated in AISC codes in recent decades were based on work by Julian and Lawrence.

Randal Wood, a scientific officer at the Building Research Station in England, published a paper (Wood 1958) titled "The Stability of Tall Buildings" that is discussed in King (2008). Wood's paper describes the fundamentals of stability of buildings, with rigid beam–column joints explaining the basis of current practice in the United Kingdom. Wood's historic paper is cited by King as being "valuable reading for all today's designers of medium rise buildings."

At the time Wood wrote the paper (1958), the design of medium-rise buildings in England was moving toward achieving continuity in steel framing to reduce gravity load deflections. This allowed beams and columns to act as frames in resisting lateral forces and, thus, affording stability. Also at that time, infill cladding was predominantly brickwork, usually providing significant resistance to the sideway. Furthermore, the use of plastic design, intended to increase the economy of steelwork, was evolving. Wood clearly recognized the problems associated with instability as follows:

Instability of individual stanchions (columns)

(a) Lateral torsional stability of columns and beams;

(b) Local crinkling (buckling) of flanges;

(c) Instability because of limitation of the materials (and joints), that is, the moment of resistance eventually falls off; and

(d) Frame instability.

Wood carried out racking tests on model steel frames three and four stories high. The models demonstrated a significant reduction in critical buckling loads.

Wood established elastic critical values for simple unsymmetrical beam and column structures for no-sway and unrestricted sidesway conditions based on stability functions established by Livesley and Chandler (1956) [Figure 2-25(a, b)]

Figure 2-25(a) indicates that, for a no-sway condition, P_{crit}/P_E values range from 0.0 (hinge each end) to 2.0 (fixed each end). Figure 2-25(b) indicates that for an unrestricted sideways condition P_{crit}/E values range from 0 (hinge each end) to 1.0 (fixed each end), illustrating the significant difference between no-sway and unrestricted sideway conditions. These charts allowed effective length factors to be selected (Wood 1974).

Wood recognized that in a frame, there is "no such thing as an individual stanchion (column)," rather "it has always been "frame" collapse.

Wood also investigated the effects of plasticity which, at that time, had not yet been looked into in the United Kingdom. Wood showed, both from tests and from an analytical approach, that when plastic hinges start to occur, the frame stability deteriorates.

Thus, Wood proceeded to establish a method that could represent the elastoplastic behavior of frames. Wood considered the effects of gravity load–causing plastic hinges at the ends of the beam and pondered that, if this occurs on all floors effectively, the columns act as free cantilevers, which he called a tendency toward "conversion to chimneys."

As lateral loading is increased to a level where all joints form a plastic hinge, a mechanism occurs such that the deteriorated critical load is zero. Even considering that if four of five joints act as plastic hinges, out of a total of 10 joints, significant deterioration occurs.

Based on a consideration of an analysis on three example frames, Wood concurred with a suggested approach by W. Merchant, derived from Rankine's Strut formula, that the following intuitive relationship would be an appropriate representation of the overall load factor for different frames:

$$\frac{1}{\lambda} = \frac{1}{\lambda_p} + \frac{1}{\lambda_c} \tag{2-48}$$

where

λ_p = Load factor for an ideal plastic collapse,
λ_c = Elastic critical load, and
λ = Actual collapse load factor.

The aforementioned formula is known as the Merchant–Rankine formula. It should be noted that Wood's approach and concerns were mainly related to the

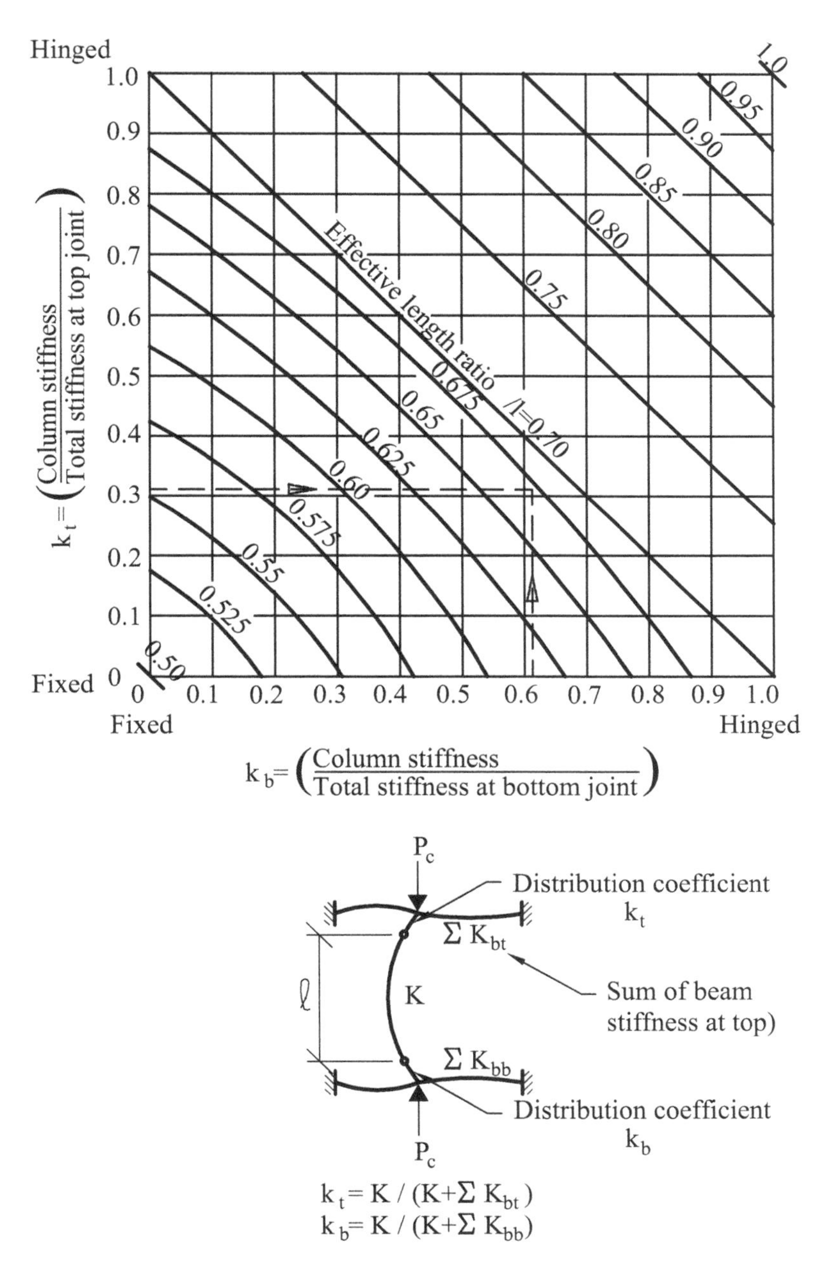

Figure 2-25.(a) Effective length ratios for single column with restraining beams and rigid joints and no sway.

Source: Reproduced from Wood (1958), courtesy of the British Research Establishment.

Note: May not be current practice.

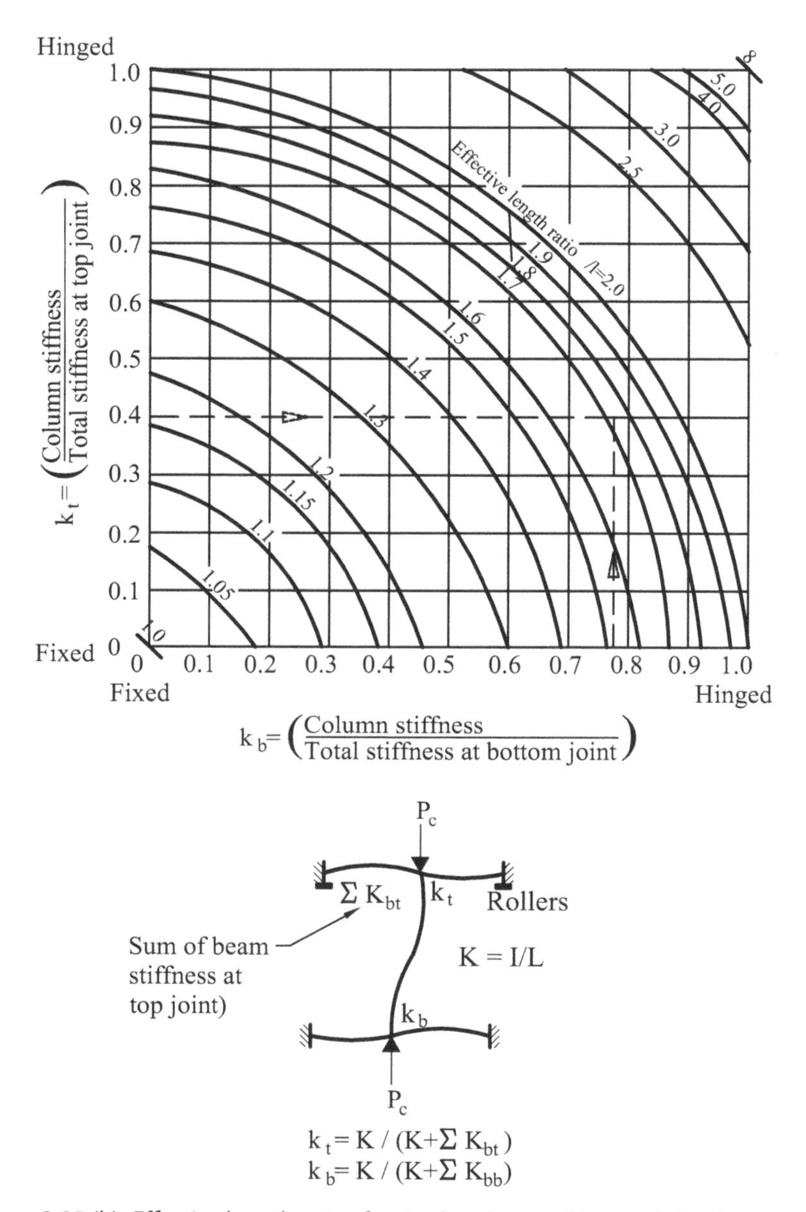

Figure 2-25.(b) Effective length ratios for single column with restraining beams and rigid joints and no restraint against side sway.

types of construction at that time with plastic hinging in beams primarily because of gravity loads and lateral wind forces. Wood warned about the deteriorating effects of plastic hinges and advised that unless bracing is introduced, "wind bents," that is, rigidly connected moment frames, should be provided.

As mentioned by King (2008), construction practice moved toward achieving rigidly jointed frames considering the implications of Wood's research.

It is very important to note that, although structures requiring to be designed for wind forces essentially behave elastically, structures subjected to seismic forces can induce a significant yielding of beam–column joints. As lateral forces increase, beam–column joints start to yield such that lateral deformations increase and the frame stiffness reduces, which can lead to instability. With reference to Johnston (1976) who cited tests by Beedle, Lu, and Lim in 1969, "all tests show conclusively that unbraced frames are likely to fail through instability before the formation of a plastic mechanism." Johnston (1976) listed the factors stated by Beedle Lu and Lim that influence frame instability. Adding to those, listed by Wood, these include the following: (Johnston 1976, pp. 427 and 428).

1. Effect of changes in geometry,
2. Changes of length from axial strain,
3. Influence of axial force on member stiffness (which can be accounted for by using stability functions),
4. Nonlinear stress–strain relationships,
5. Residual stresses,
6. Initial imperfections,
7. Nonproportionality of loading,
8. Shearing strains,
9. Strain reversal,
10. Axial force decreasing plastic moment capacity,
11. Spread of the inelastic zone as yielding (at hinges) progresses,
12. Local buckling of members,
13. Lateral torsional buckling of members,
14. Out-of-plane torsional buckling of frames,
15. Variable repeated loading,
16. Overlapping of members at joints,
17. Composite action with floor slabs, and
18. High applied axial loads because of uncertainties of earthquakes and wind loads.

This can be represented by curves obtained from tests on single-story single bay frames [Figure 2-26, derived from Johnston (1976, p. 427)]. It can be seen that the frame capacity is significantly reduced. A second-order analysis is necessary to address the effects of axial force on member stiffness, commonly known as the "P-delta effects." Additional moments and shears occur because of the gravity loads applied to the lateral deformations that occur primarily on account of lateral forces. An increase in load increases the stresses until

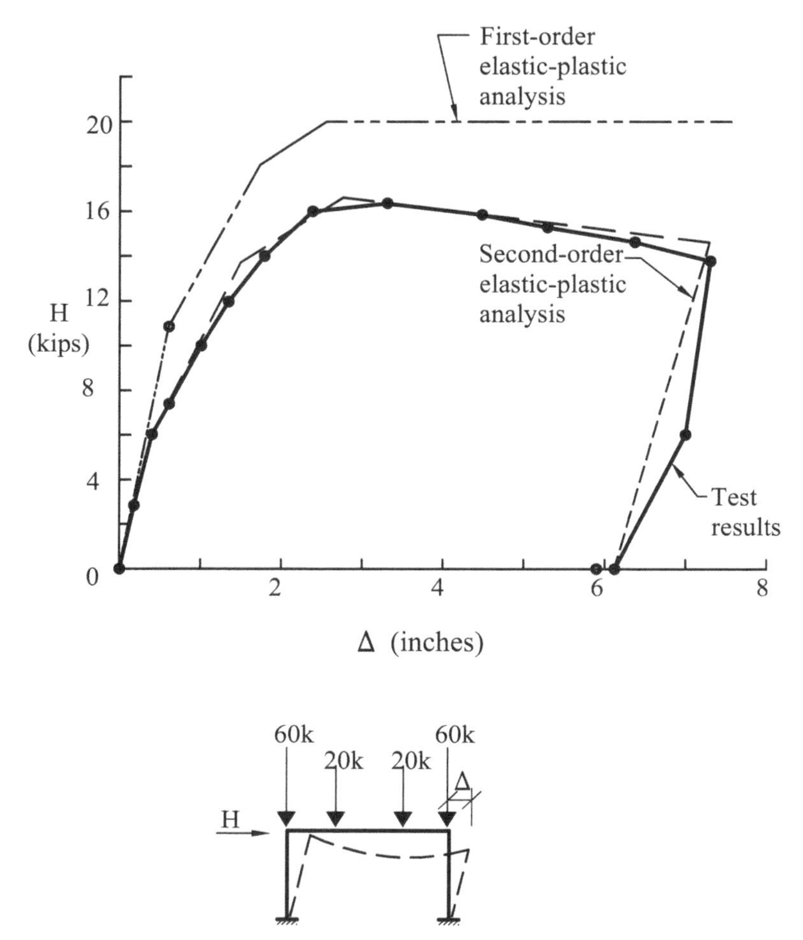

Figure 2-26. Observed and predicted load–deflection relationships. Rigid frame subjected to combined gravity and lateral loads.

Source: Reproduced from Figure 15.8 in Johnston (1976, p. 427), courtesy of Wiley.

yielding occurs. Eventually, progressive yielding and lateral displacement along with axial loads in the members reduce the frame stiffness such that the frame becomes unstable.

The subject of global instability of framed structures, subject to yielding, in the opinion of the writer, is still a subject not well understood and not readily quantifiable. This particularly occurs with frames undergoing seismic events in which numerous joints, subjected to cyclic action that can yield, then degrade because of local buckling and "P-delta" effects from gravity loads. Yet, as degradation tends to cause leaning leading to instability, forces tend to reverse with cyclic behavior. Further research to improve the understanding of frame stability beyond that of past researchers appears warranted.

2.9 LOCAL BUCKLING, BUCKLING OF PLATES SUBJECT TO COMPRESSION

Local buckling, including buckling of plates, subject to compression, is much influenced by boundary conditions that can vary significantly.

An excellent and thorough discussion on the development of the understanding of local buckling of plates is given in Bleich (1952, pp. 302–357). The following is a brief outline of the discussions by Bleich.

Investigations on the stability of plates that partly addressed local buckling issues took place in the late nineteenth century. G. H. Bryan in 1891 applied the energy criterion to derive a "buckling stability solution," based on elastic behavior, for a rectangular plate, simply supported along all edges.

The following relationship was established for elastic behavior (Johnston 1976, p. 83; Gerard 1962, p. 40):

$$\sigma = K\left(\frac{\pi^2 E}{12\left(1-\nu^2\right)\left(b/t\right)^2}\right) \tag{2-49}$$

where

σ = Critical elastic stress,
k = Buckling coefficient,
b/t = Width-thickness ratio,
ν = Elastic Poisson's ratio, and
k = Buckling coefficient determined from theoretical analysis.

Where plates are in constant compression and both nonloaded edges are simply supported, $k = 4.00$. Where plates are in constant compression with one edge free and the other nonloaded edge simply supported, $k = 0.425$.

Timoshenko developed more extensive solutions to the buckling of plates for various conditions subject to elastic behavior, circa 1906. These are discussed in Timoshenko and Gere (1961). Contributions were also made by Hans Reissner. F. Bleich extended the theory on flat plates to the inelastic range in 1924. Solutions developed by P. P. Bijlaard in 1940 showed good agreement with tests on aluminum plates supported on four sides carried out by C. F. Kollbrunner. Bleich (1952) gives credit to E. Lundquist, E. Stowell, and E. Schuette, who carried out extensive analysis on the stability of plate assemblies using the moment distribution method. Large-scale tests were carried out between 1935 and 1946 by Kollbrunner in Zurich, Austria. Tests were also carried out on local buckling by G. Gerard in 1946 in the United States, primarily to establish the effective modulus.

Gerard utilized the secant modulus E_s and established, for the elastic to the inelastic range, the following relationship (Bleich 1952, p. 352):

$$\frac{\sigma_c}{E_s} = K\left(\frac{t}{b}\right)^2 \tag{2-50}$$

This showed good agreement with tests on the channel and Z sections. However, this did not include all boundary conditions.

Equations for plates of finite length, simply supported at the edges, were established by Gerard and Becker (1957) (Dowswell 2010).

It should be understood that, whereas the capacity of a column terminates when buckling occurs, in the case of plates, tensile stresses, perpendicular to the direction of load, tend to stiffen the plate. This is illustrated in Figure 2-27(a) (derived from Gerard 1962, p. 45) and Figure 2-27(b) (derived from Gerard 1962, p. 46). Figure 2-27(a) indicates compressive forces on two simply supported opposite sides. The compressive stresses are no longer uniform, decreasing to the center because of the deformations that occur in the panel. The unloaded, simply supported constrained and straight sides exert tensile stresses at yield near the corners of the plates. The transverse tensile stresses provide stiffening that appreciably reduces out-of-plane deformation such that the buckling load is significantly increased above that which would occur without constraint at the sides.

Stowell developed a theory that took account of various boundary conditions. For a simply supported, flat, rectangular inelastic plate, Stowell derived the following equation (Gerard 1962, p. 52):

$$\sigma_{cr} = \frac{\pi^2 K_c \eta_c E}{12\left(1-\nu^2\right)}\left(\frac{t}{b}\right)^2 \tag{2-51}$$

where

$\nu =$ Elastic Poisson's ratio,

$K_c =$ Coefficient dependent on the plate geometry, and

$\eta_c = \dfrac{\sigma_{cr}\ \text{(plastic)}}{\sigma_{cr}\ \text{(elastic)}}$

Bleich proposed, for inelastic buckling, the following (Johnston 1976, p. 85):

$$\sigma = K\left[\frac{\pi^2 E\sqrt{\eta}}{12\left(1-\nu^2\right)\left(b/t\right)^2}\right] \tag{2-52}$$

where

$\eta = \dfrac{E_t}{E}$, and

$E_t =$ Tangent Modulus

The aforementioned procedure, or a similar procedure, is used to establish the width-to-thickness ratios of flanges and height-to-thickness ratios of webs of elements such that local buckling does not occur before buckling of the whole member (global buckling) or yielding of the member dependent on demand. AISC provides criteria for the following:

- Noncompact sections,
- Compact sections, and
- Sections subject to yielding (seismic events).

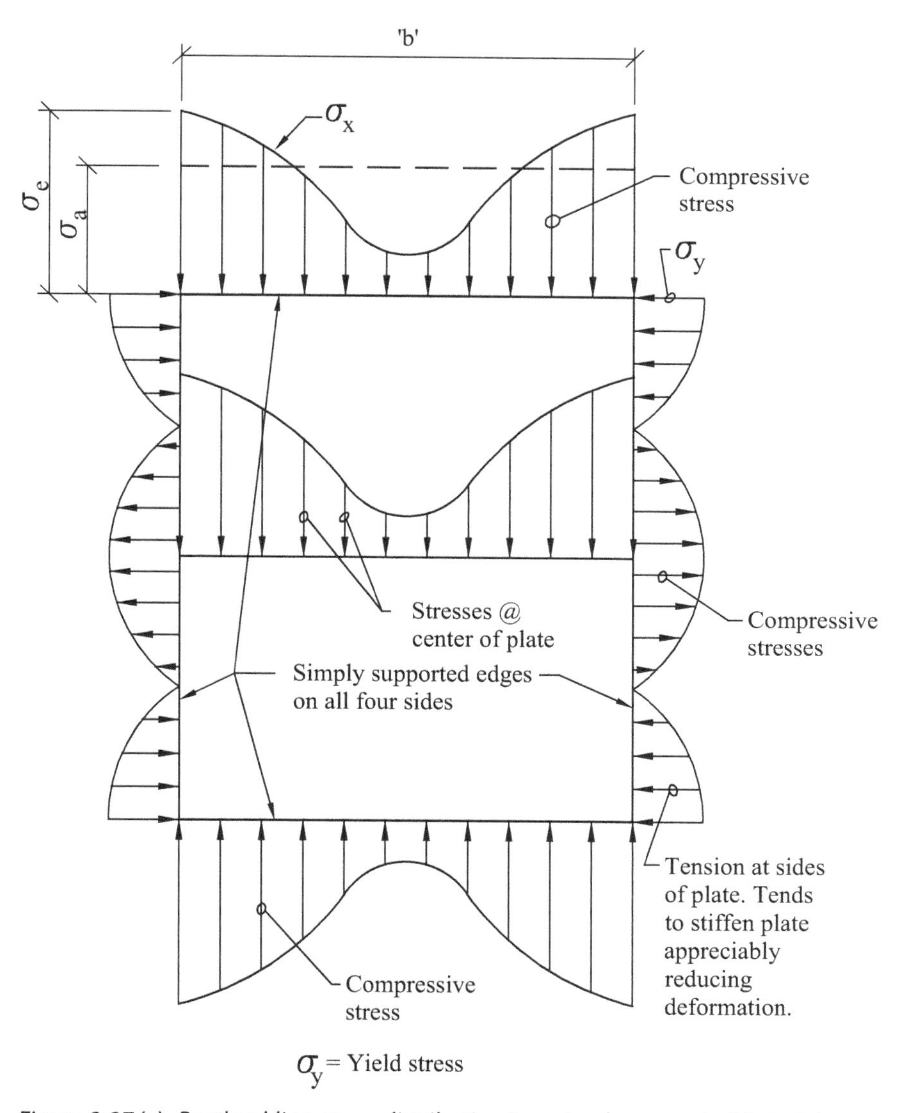

Figure 2-27.(a) Postbuckling stress distribution in a simply supported flat plate.

Source: Reproduced from Gerard (1962) page 45, Courtesy of the McGraw-Hill Global Education Holdings, LLC.

This is carried out by providing criteria for b/t ratios.

From the aforementioned equations and assuming that the critical stress is at yield (σ_y), the following relationship is obtained (Johnston 1976, p. 89):

$$\frac{b}{t} \leq \sqrt{\frac{k\pi^2 E}{12(1-n^2)\sigma_Y}} \tag{2-53}$$

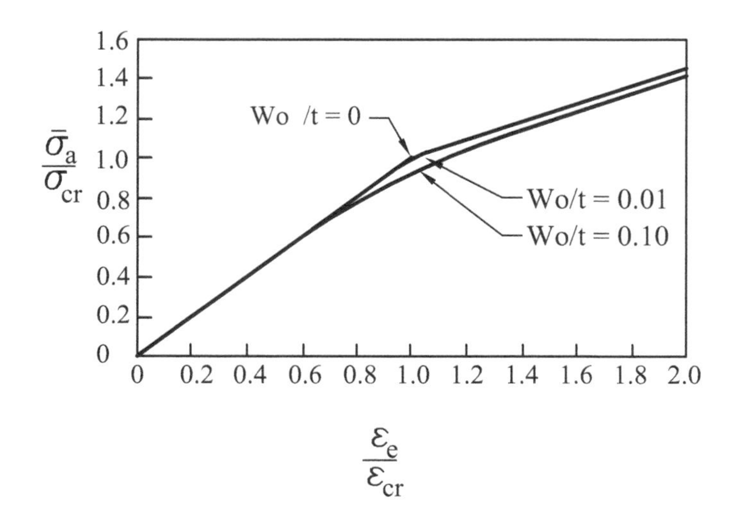

$\bar{\sigma}_a$	=	Average compressive stress
σ_{cr}	=	Buckling stress
ε_e	=	Edge compressive strain
ε_{cr}	=	Buckling strain
Wo	=	Intitial imperfection
t	=	Thickness

Figure 2-27.(b) Postbuckling load-carrying ability of compressed plates.

Source: Reproduced from Gerard (1962) page 46, Courtesy of the McGraw-Hill Global Education Holdings, LLC.

Long plates, subject to compression in the longitudinal direction tend to buckle in waves, as illustrated in Figure 2-28.

The boundary conditions can significantly affect the potential for local buckling. Stowell and Lundquist, in 1942, established charts to represent the restraint of flanges, subject to global bending stresses, of "*I*" sections by the web. As expected, the deeper the depth of the section, the less restraint by the web is afforded to the flanges. On the contrary, flanges can afford restraint to web buckling because of shear. This is also illustrated in Figure 2-29 (from Johnston 1976, p. 88) for I section. The coefficient, K_w may be used in Equation (2-57). It can be seen in Figure 2-29 that narrowing the flanges causes the web to buckle earlier because this results in less restraint being provided. On yielding of the member, strain hardening occurs, with increased distortion.

An example of buckling in the plastic region is given in Figure 2-30 [Figure 1.7c from Maranian (2010) Steel Failure document], which shows a diagonal brace failure with significant local buckling at the mid position of the brace. This

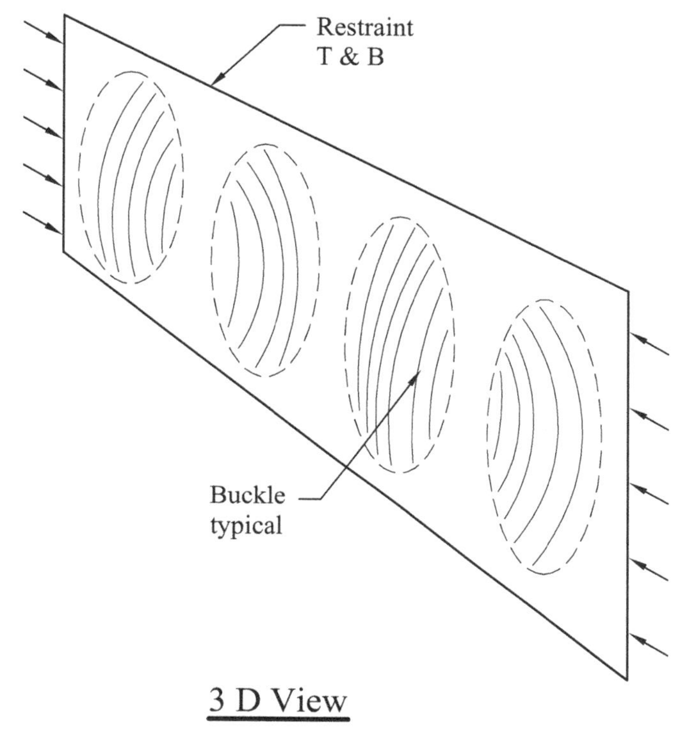

Figure 2-28. Buckling in waves of long plate subject to longitudinal compression.

occurred in a three-storied building in Los Angeles during the 1994 Northridge Earthquake. It appears that the local buckling along with axial tension stresses caused the failure.

Figure 2-31(a to c) indicate local buckling of flanges during a beam-to-column test on a bolted flange plate moment connection. Elongated holes were provided in the flanges to prevent net section failure. At each cycle, the flanges are yielded alternately in tension and compression with an increased permanent deformation in plane at each stage. The compression also starts to cause out-of-plane deformations. Thus, when the compression phase occurs, there is appreciable bending because of the eccentricities caused by the out-of-plane deformation. The tension phase tends to straighten out the member, but residual deformations owing to yielding remain and subsequent cycles cause increased out-of-plane deformations. As can be seen, significant local buckling occurred in a wavelike manner. The local buckling is indicative of postelastic local buckling. The test satisfied the requirements of AISC Seismic (AISC 2002), Appendix S for Steel Moment Frame connections, with failure eventually occurring as a result of fracture in one flange because of local cycle fatigue accentuated by the local buckling.

The cyclic behavior at yield causes fatigue. Because there are only a relatively few cycles, the phenomena is known as *low cycle fatigue*, eventually leading to fracture.

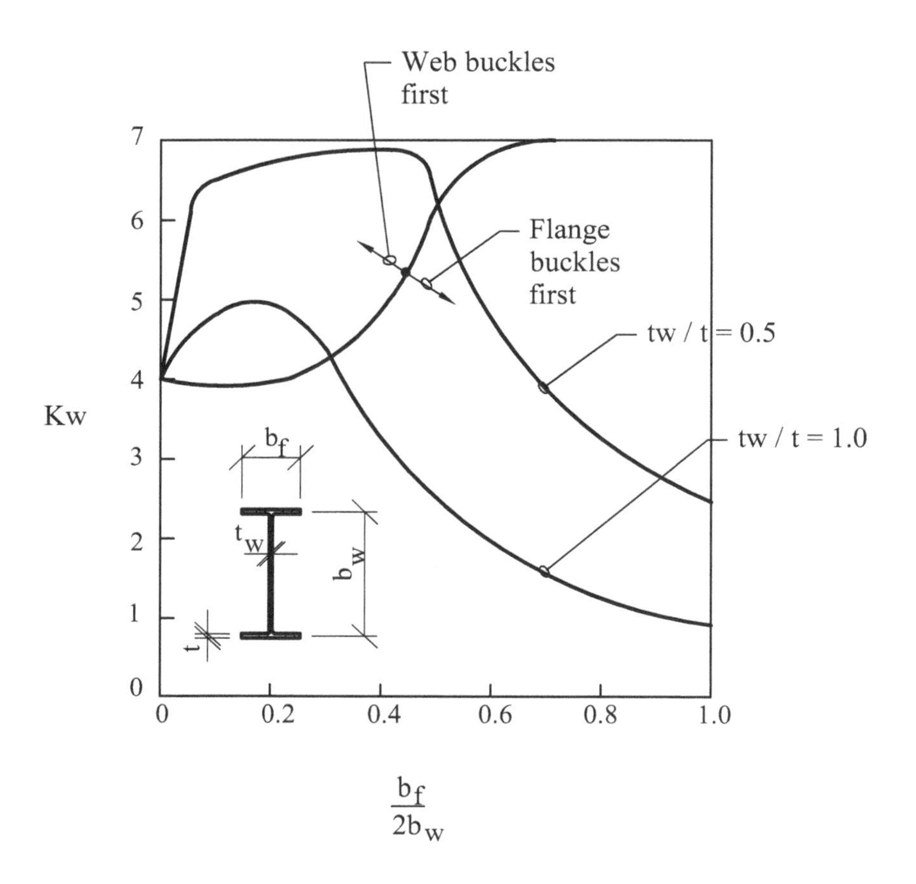

Figure 2-29. Plate-buckling coefficient K_w for wide-flange columns (Ref. 4.4).
Source: Reproduced from Johnston (1976, p. 88), courtesy of Wiley.

Cyclic tests carried out by Bertero and Popov (1965) on cantilevered 4 in. deep wide flange members in the elastic and inelastic range caused significant local buckling and torsional displacements. Bertero and Popov advised that the problem of preventing local buckling is considerably more important, which concerns the low cycle fatigue endurance of the material itself. Also, see Chapter 4, Section 4.2, regarding steel properties and Chapter 2, Section 2.13, regarding Nonlinear Buckling.

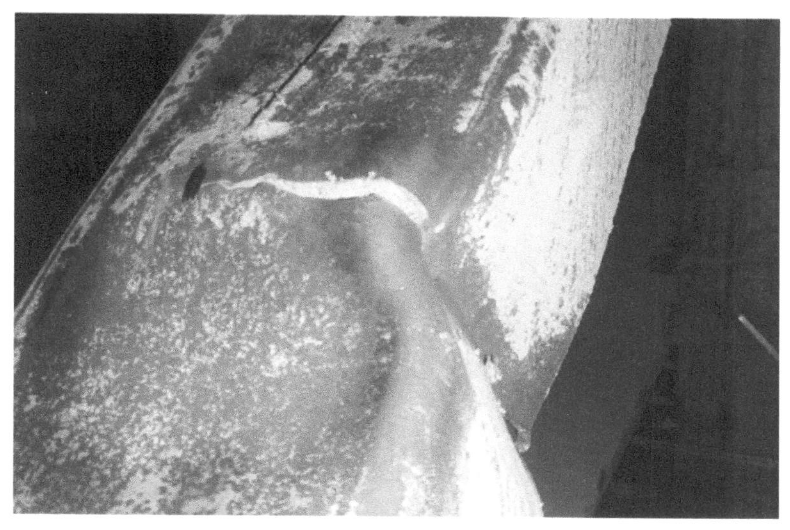

Figure 2-30. Local buckling of steel tube brace member at the center of brace Northridge Earthquake, Los Angeles, 1994.

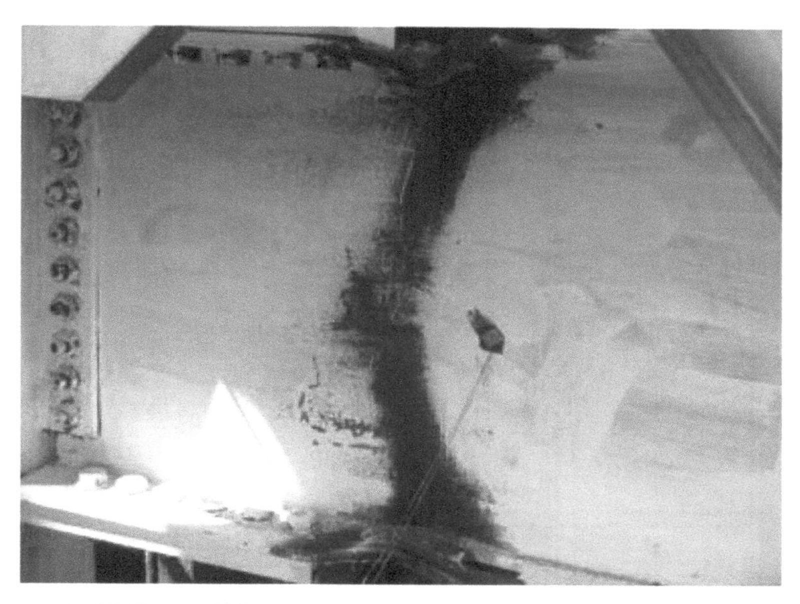

Figure 2-31.(a) Test on BFP connection with elongated holes, circa 2008.

Figure 2-31.(b) Test on BFP connection with elongated holes test, circa 2008: Top flange buckling after achieving the required test performance.

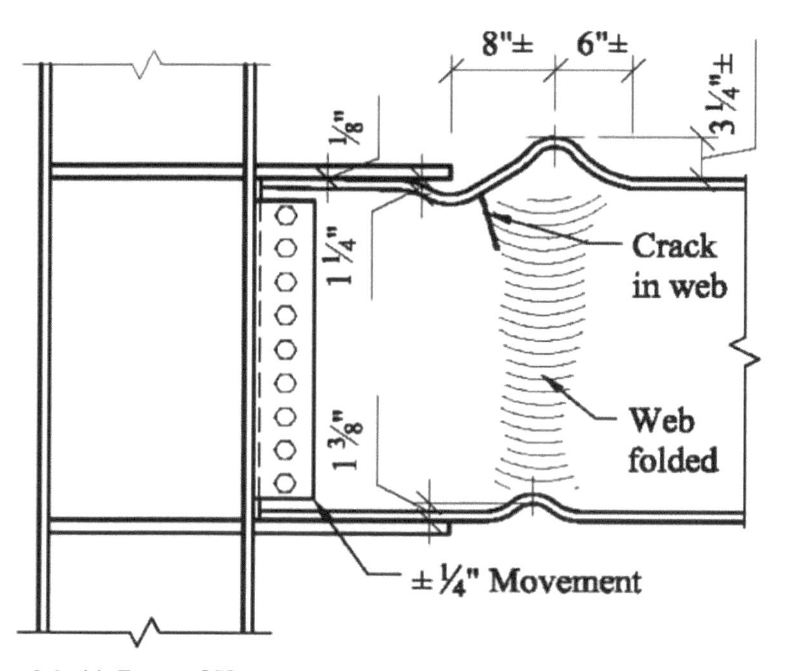

Figure 2-31.(c) Test on BFP connection with elongated holes, circa 2008: Elevation near side (at end of test).

2.10 BUCKLING OWING TO SHEAR

Plates subject to edge shear stresses shown in Figure 2-32 [from Figure 4.12, Johnston (1976, p. 104)] induce tension and compressive principal stresses typically inclined at a 45 degree angle. Based on a similar approach to plate theory, critical elastic shear stress (Bleich 1952, p. 393)

$$\tau_c = \frac{K_s \pi^2 E}{12(1-\nu^2)(b/t)^2} \tag{2-54}$$

where K_s is the critical shear stress coefficient subject to boundary conditions.

Solutions, for different support conditions, were established by several researchers, including Timoshenko, S. Bergmann and H. Reissner, E. Seydel, R. V. Southwell and S. Skan, S. Iguchi, D. M. A. Leggett, I. T. Cook, and K. C. Rockey. The curves, for different support conditions, are indicated in Figure 2-32 [from Figure 4.12 of Johnston (1976, p. 104)].

With regard to *plate girders*, significant postbuckling capacity strength, beyond the elastic buckling capacity, is achieved by tension field action that was developed by H. Wagner in the 1930s. His interests were mainly in aircraft. Further analytical and experimental studies were carried out by K. Basler and B. Thürlimann in the 1950s, considering the postbuckling behavior of web subjected to shear and bending. They developed a formula that assumed that the flanges were too flexible to resist the tension field forces. Tests were carried out by M. A. D'Apice, D. J. Fielding, and P. D. Cooper, circa 1966. Cooper then developed simplified and conservative procedures combining the critical elastic shear stress with the subpanel tension field capacities. C. Gaylord, circa 1972, then T. Fujii, and later, A. Selberg showed that the formula by Basler and Thürlimann overestimated the shear capacity. Gaylord, Fujii, and Selberg, using a limited tension band across the web, presented a modified formula. D. M. Porter, K. C. Rockey, and H. R. Evans, circa 1974, also developed procedures, using an iterative approach, that were in good agreement with tests carried out by Rockey, circa 1974 (Johnston 1976, p. 152 and 174).

Investigations, carried out by C. Chern and A. Ostapenko, circa 1970, derived an iterative approach solved by computer analysis. M. Herzog, who analyzed 19 tests, developed simplified formulas circa 1973, which were less accurate than those developed by Rockey et al. and Chern and Ostapenko (Johnston 1976, p. 176).

Essentially, the wider the spacing of the vertical web stiffeners, the lower the benefit of tension field action, such that with an increased spacing of vertical web stiffeners, the shear strength eventually reduces to that based on the critical elastic shear stress. The tension field is also influenced by the bending stiffness of the flanges. Thus, the size of the flange plates affects the tension field, as illustrated in Figure 2-33(a), where the stiffer and stronger flange plates result in a steeper incline in the tension field compared with weaker flange plates, where the

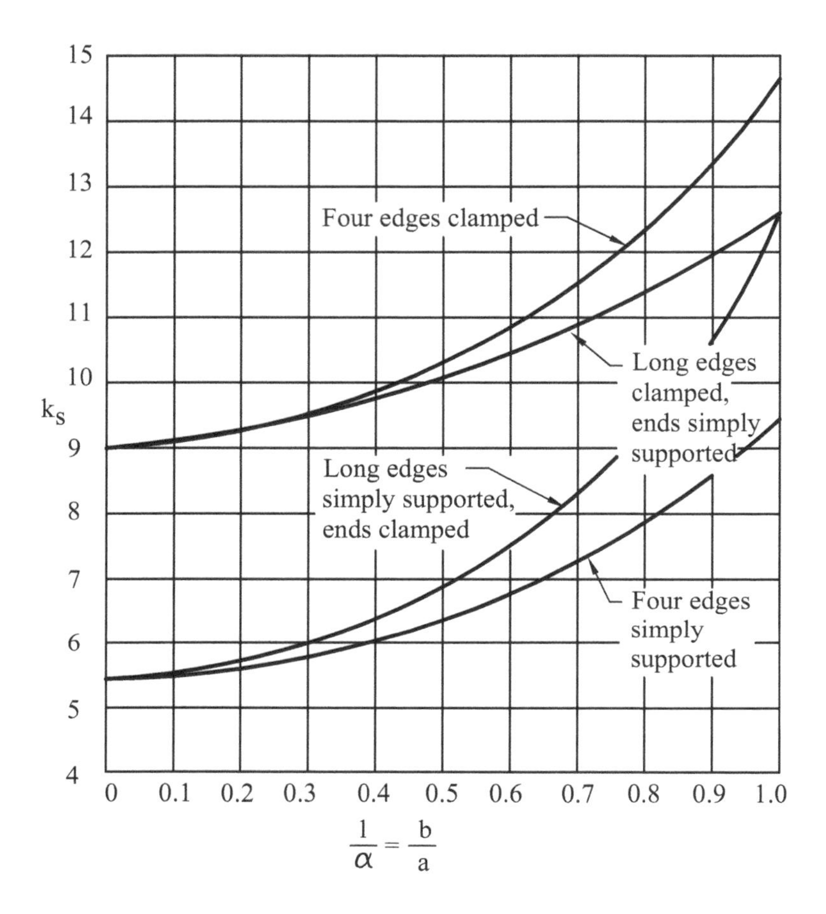

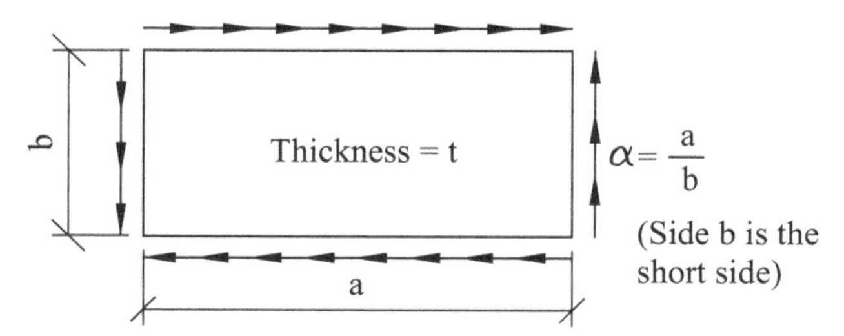

Figure 2-32. Buckling coefficient for plates in pure shear.

Source: Reproduced from Johnston (1976, p. 104), courtesy of Wiley.

tension field is much shallower. As the shear is increased, the web tends to yield and plastic hinges occur in the flanges. These plastic hinges occur further away from the vertical stiffeners for the stiffer and stronger flange plates compared with the weaker flange plates, shown in Figure 2-33(b), because of the higher bending capacities of the flanges.

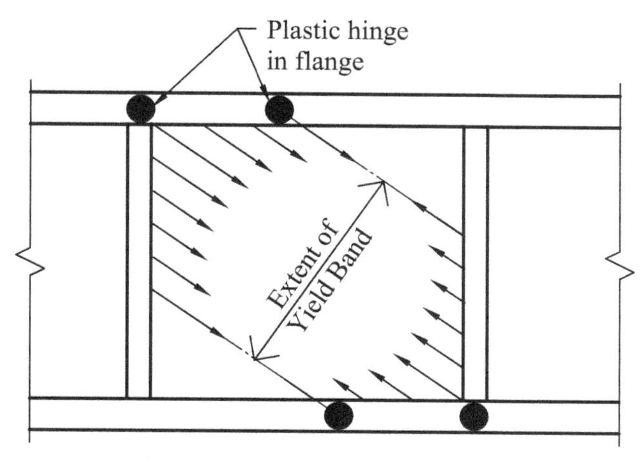

Figure 2-33.(a) Tension field action with stiffer/stronger flange plates.

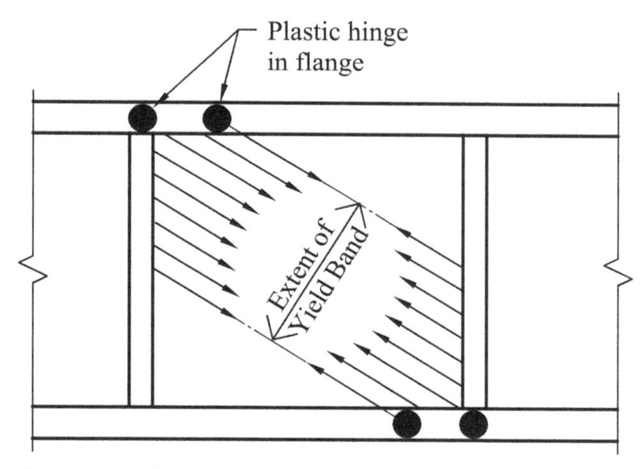

Figure 2-33.(b) Tension field action with less stiff/weaker flange plates.

Vertical stiffeners, as shown in Figure 2-34, are required to resist the vertical component of the diagonal tension. These also need to be checked for buckling. End stiffeners are also required to accommodate the inward pulling of the diagonal membrane forces. This can be achieved with a pair of stiffeners, which, along with the web between the stiffeners, can be designed as a post subject to flexural bending.

Longitudinal stiffeners, as shown in Figure 2-35, can significantly enhance the postbuckling capacity strength of plate girders by reducing the b/t ratio and increasing the critical elastic shear stress. The best location of longitudinal stiffeners to increase shear buckling capacity is at the middepth. For an idea of the effectiveness of longitudinal stiffeners for pure bending, see Section 2.12.

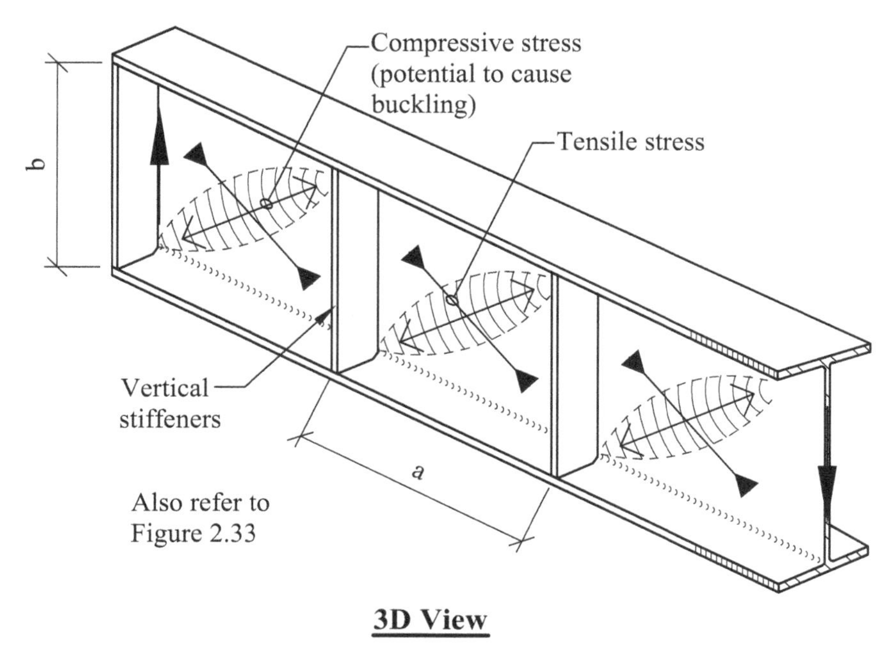

Figure 2-34. Shear buckling controlled by vertical stiffeners.

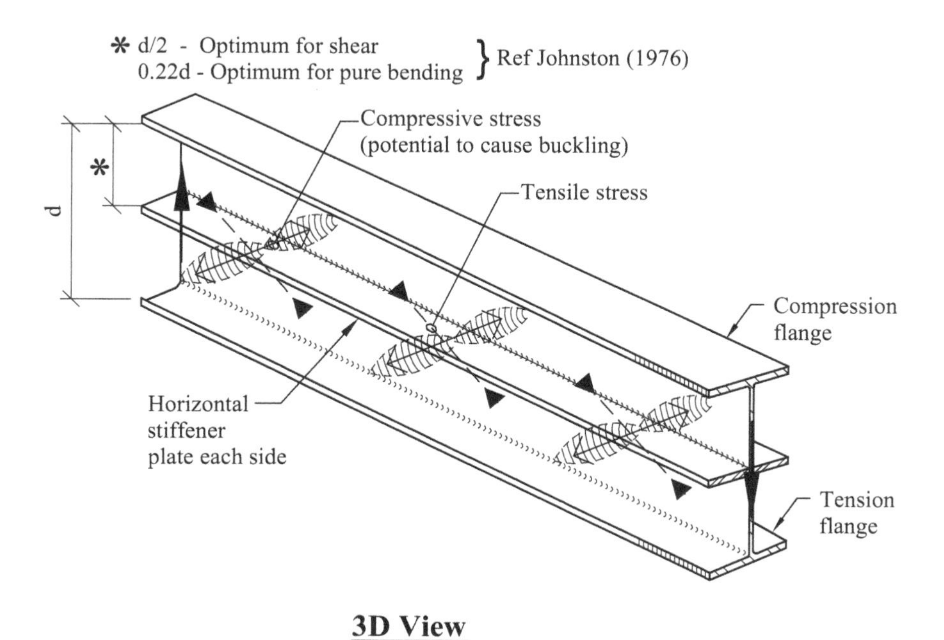

Figure 2-35. Shear buckling controlled by horizontal stiffeners.

2.11 STEEL PLATE SHEAR WALLS

In recent years, the concepts of tension field action have been adopted to develop steel plate shear walls [as lateral resisting systems in both seismic and wind areas (Berman and Bruneau 2004, AISC 2005b)]. These comprise infill steel plates with columns and beams acting as vertical and horizontal boundary elements (HBEs). In seismic areas, the infill plates are intended to yield, providing energy dissipation. The columns, which act as vertical boundary elements (VBEs) and the HBEs, are required to remain elastic while being subjected to the tension field forces imposed by the infill plate. Several tests have been carried out on this system in columns up to four stories tall (Berman and Bruneau 2004, Driver et al. 1998, Li and Tsai 2008, Qu et al. 2008]. The tests appeared to indicate good performance. Significant buckling of the infill plates occurs typically in a wavelike manner that was first predicted by H. Wagner in 1929 (Wagner 1929). Control of buckling is partly dependent on the strength (including buckling) and stiffness of the VBEs and HBEs. Wagner's analysis is utilized by Bing and Bruneau (2010) in their investigation on criteria for boundary conditions to achieve acceptable performance. The less the stiffness of HBEs, the greater the variation of tension field force distribution across the infill plate.

This system has been used in some buildings but may not yet have been subjected to significant seismic events. Questions on the behavior of this system include the shortening effects on columns, owing to gravity loads and overturning forces, along with the complex behavior resulting from higher modes.

2.12 WEB BUCKLING BECAUSE OF PURE BENDING

Plate girders, when primarily resisting bending, if deep, fail because of lateral torsional buckling or local buckling of the compression flange. The addition of longitudinal stiffener plates, according to Bleich (1952, p. 420), when located at the center line of the web, as shown in Figure 2-35, provides about a 50% increase in the buckling strength. Bleich also states that horizontal stiffener plates are less effective in the inelastic range. According to Bleich (1952, p. 420), horizontal stiffener plates were studied by Chwalla in 1936, with the plates located at 0.2 times the depth of the web from the compression flange, and a more detailed study was done by C. Massonnet in 1941. According to Johnston (1976, p. 173), the optimum location of the longitudinal stiffener plates is 0.22 times the depth of the web from the compression flange.

The complex problem of shear, combined with longitudinal compression in bridge girders with deep web plates, reinforced with transverse and longitudinal stiffener plates, was investigated by M. Milosavljevitch, circa 1947, resulting in tables being provided to carry out routine calculations.

With reference to Bleich (1952, p. 422), C. Dubas, in 1948, derived charts for plates reinforced with longitudinal stiffener plates for various aspect ratios to determine the critical shear stress factor K_s in Equation (2-36). An example of Dubas's charts is given in Figure 2-36.

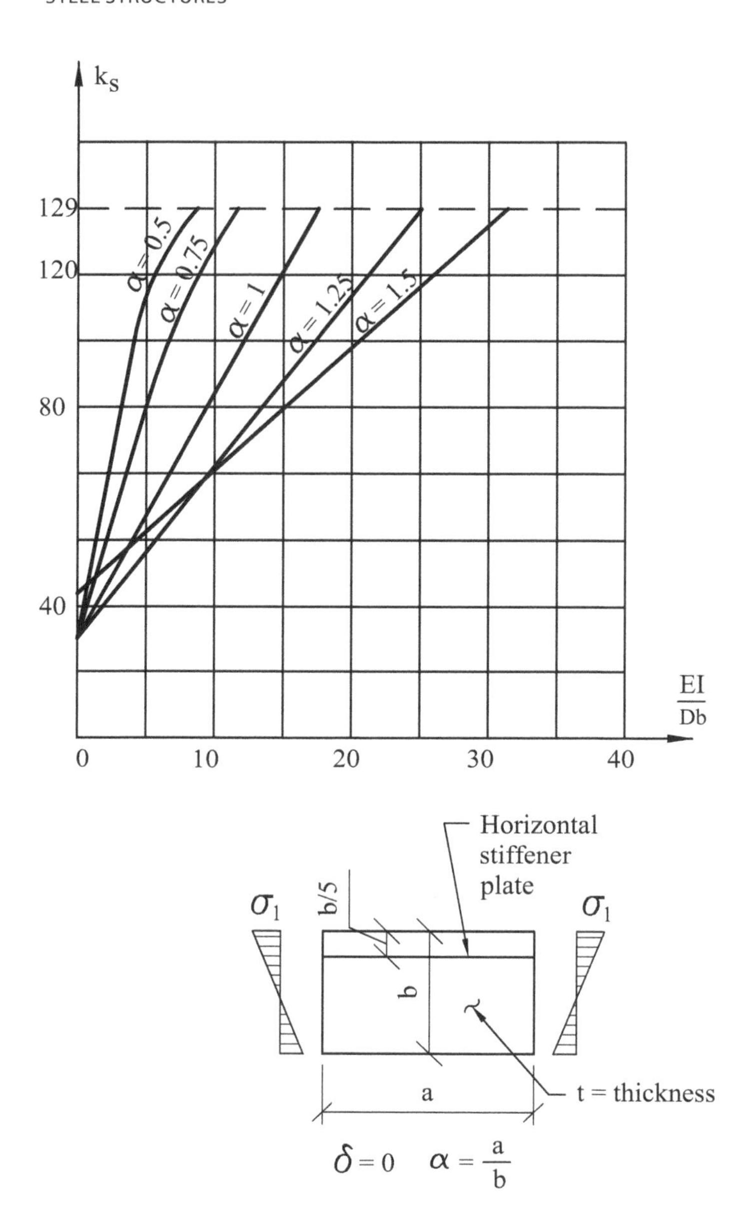

Figure 2-36. Chart by Dubas for various aspect ratios for longitudinal stiffeners b/5 from compression flange.

Source: Reproduced from Figure 218 in Bleich (1952, p. 422), courtesy of United Engineering Foundation.

2.13 NONLINEAR BUCKLING

Buckling occurs as a result of membrane forces in the cross thickness direction of the thin members/parts of the slender element. An increase in membrane compressive forces in these members/parts decreases the lateral stiffness of the membrane, eventually reducing it to zero and resulting in buckling failure. The bending moment caused by an eccentricity of the applied forces further reduces the applied load where buckling failure occurs. This is the mechanism of buckling failure of slender structures and occurs entirely in the elastic range, hence the term *elastic blocking*.

Typical elastic theories underestimate the buckling failure loads because of the reasons explained in Section 2.8. For this reason, for a structural framing member or its parts identified to be at or postbuckling regions, a nonlinear buckling analysis is required. Typical design formulas used for checking the adequacy of individual members cannot be used to understand and assess structural performance, especially because of the application of dynamic axial and out-of-plane forces resulting from earthquakes, wind(s), hurricanes, and tornadoes. This is especially true for structures designed to take advantage of ductile behavior of its parts or joints as it changes the boundary conditions of the element or part prone to buckling. The same is true for moderately slender members/braces of the structural frame, where compression loads approach the yield stresses of the member/brace, resulting in nonlinear buckling behavior.

In conclusion, other than a preliminary analysis or designing a structure based on very conservative assumptions, to design a reliable and efficient structural framing system, nonlinear buckling analysis should be considered as an essential part of the design process.

References

AISC (American Institute of Steel Construction). 1948. *Steel construction*. Chicago: AISC.
AISC. 1963. *Specification for the design, fabrication and erection of steel for buildings*. 6th ed. New York: AISC.
AISC. 1978. *Specification for the design, fabrication and erection of structural steel for buildings*. Chicago: AISC.
AISC. 2002. *AISC Seismic provisions for structural steel buildings*. Chicago: AISC.
AISC. 2005a. *Prequalified connections for special and intermediate steel moment frames for seismic applications*. ANSI/AISC 358. Chicago: AISC.
AISC. 2005b. *Seismic design provisions for structural steel buildings*. ANSI/AISC 341 Seismic. Chicago: AISC.
AISC. 2016a. *Prequalified connections for special and intermediate steel moment frames for seismic applications*. ANSI/AISC 358. Chicago: AISC.
AISC. 2016b. *Seismic design provisions for structural steel buildings*. ANSI/AISC 341 Seismic. Chicago: AISC.
AISC. 2016c. *Specification for structural steel buildings*. ANSI/AISC 360. Chicago: AISC.
Berman, J. W., and M. Bruneau. 2004. "Steel plate shear walls are not plate girders." *Eng. J. Am. Inst. Steel Constr.* 41 (3): 95–106.

Bertero, V. V., and E. P. Popov. 1965. "Effect of large alternating strains of steel beams." *J. Struct. Div.* 91 (1): 1–12.
Bjorhovde, R. 1988. "Columns: From theory to practice." *Eng. J. Am. Inst. Steel Constr.* 25 (11): 21–34.
Bleich, F. 1952. *Buckling strength of metal structures.* New York: McGraw Hill.
BSI (British Standards Institution). 1969. *Specification for the use of structural steel in building.* British Standard 449, Part 2. London: BSI.
Clark, J. W., and H. N. Hill. 1960. "Lateral buckling of beams." *J. Struct. Div.* 86 (ST7): 175–196.
Dowswell, B. 2010. "Stiffener requirements to prevent edge buckling." *Eng. J. Am. Inst. Steel Constr.* 47 (2): 101–108.
Driver, R. G., G. K. Kulakm, D. J. L. Kennedy, and A. E. Elwi. 1998. "Cyclic test of a four-story steel plate shear wall." *J. Struct. Eng.* 124 (2): 112–120.
Dumonteil, P. 1999. "Historical note on K – factor equations." *Eng. J. Am. Inst. Steel Constr.* 36: 102–103.
Gerard, G. 1962. *Introduction to structural stability theory.* New York: McGraw-Hill.
Gerard, G. and H. Becker. 1957. *Handbook of structural stability, III: Buckling of curved plates and shells.* NACA TN3783. Washington, DC: National Advisory Committee for Aeronautics.
Horne, M. R. 1963. *The plastic design of columns.* London: British Constructional Steelwork Association.
Johnston, B. V. 1976. *Guide to stability design criteria for metal structures structural stability research council.* Hoboken, NJ: Wiley.
King, C. 2008. "R.H. Wood's, 'The stability of tall buildings': A contemporary review." *Struct. Build.* 161 (SB5): 243–246.
Kirby, P. A., and D. A. Nethercot. 1979. *Design for structural stability.* New York: Halsted Press.
Li, C. S., and K. C. Tsai. 2008. "Experimental response of four 2-story narrow steel plate shear walls." In *Proc., 2008 Structures Congress.* Vancouver, BC. Reston, VA: ASCE. https://ascelibrary.org/doi/10.1061/41016%28314%29101
Livesley, R. K., and P. B. Chandler. 1956. *Stability functions for structural frameworks.* Manchester, UK: Manchester Univ. Press.
Maranian, P. J. 2010. *Reducing brittle and fatigue failures in steel structures.* Reston, VA: ASCE.
McGuire, W. 1968. *Steel structures.* Englewood Cliffs, NJ: Prentice-Hall.
Nethercot, D. A., and N. S. Trahair. 1976. "Lateral buckling approximations for elastic beams." *Struct. Eng.* 54 (6): 197–204.
Pincus, G. 1964. "On the lateral support of inelastic columns." *Eng. J. Am. Inst. Steel Constr.* 1 (4): 113–115.
Popov, E. P. 1952. *Mechanics of materials*, 342–366. Hoboken, NJ: Prentice-Hall.
Popov, E. P. 1959. *Mechanics of materials.* 9th ed. Hoboken, NJ: Prentice-Hall.
Qu, B., M. Bruneau, C. H. Lin, and K. C. Tsai. 2008. "Testing of full scale two-story steel shear walls with RBS connections and composite floor." *J. Struct. Eng.* 134 (3): 364–373.
Salvadori, M. G. 1955. "Lateral buckling of I-beams." *ASCE Trans.* 120: 1165–1177.
Shanley, R. V. 1947. "Inelastic column theory." *J. Aeronaut. Sci.* 14 (5): 261–267.
Steel Designer's Manual. 1972. *Constructional steel research and development organization.* 4th ed., 811–813. London: Crosby Lockwood Staples.
Timoshenko, S. P. 1953. *History of strength of materials.* New York: Dover.
Timoshenko, S. P., and J. M. Gere. 1961. *Theory of elastic stability.* 2nd ed. New York: McGraw-Hill.

Wagner, H. 1929. Ebene Blechträgenmit Sehrdünnem Stegblech, Zeitschriftfür Flugtechnik and Motorluftschiffahrt, Vol. 20.

Wilkerson, S. M. 2005. “Improved coefficients for elastic lateral–torsional buckling.” In *Proc., Structural Dynamics and Materials Conf.* AIAA 2005-2352. Austin, TX: American Institute of Aeronautics and Astronautics.

Winter, G. 1960. “Lateral bracing of columns and beams.” *ASCE Trans.* 125: 809–825.

Wong, E., and R. G. Driver. 2010. “Critical evaluation of equivalent moment factor procedures for laterally unsupported beams.” *Eng. J. Am. Inst. Steel Constr.* 47 (1): 1–20.

Wood, R. H. 1958. Discussion by various contributors regarding “The Stability of Tall Buildings”. In *Proc., Institution of Civil Engineers*, 502–522. Also published in Structures and Buildings, *Proc. Inst. Civil Engineers.* 161 (SB5): 258–263.

Wood, R. H. 1974. *Effective length of columns in multi-story buildings.* CP 85/74. Building Research Establishment, Garston, Watford, U.K.

Yura, J. A. 1993. “Is your structure suitably braced?”. April 6–7, 1993 Conference - Milwaukee, Wisconsin; Structural Stability Research Council.

Yura, J. A. 1999. *Bracing for stability* (revised). Chicago: Structural Stability Research Council, American Institute of Steel Construction. Also included in Yura and Helwig.

Yura, J. A., and T. A. Helwig. 2001. *Bracing for stability.* Chicago: Structural Stability Research Council, American Institute of Steel Construction.

CHAPTER 3

Considerations Relating to External and Internal Forces

In general, the requirements for external and internal forces are well defined in codes and design manuals. The significant development of computer analysis software programs, carried out over the last six decades, has appreciably improved the ability to analyze complex structures and assess second-order effects. Postprocessing software programs, for elemental design, have further supplanted many of the calculations previously carried out by hand.

3.1 GRAVITY LOADS AND ANALYSIS

Considerations include the following:

(a) Accuracy of determining the loads;

(b) Analysis to include second-order effects: This includes the consideration of fixity and continuity of connections, inducing moments in members, in addition to compressive forces (e.g., in trusses and braced frames);

(c) Proper consideration of unbraced length: This includes accounting for restraint conditions;

(d) Accurate representation of lateral restraint/bracing locations in the analysis;

(e) Consideration of load application at the top versus center or bottom of member. This is discussed in Chapter 2, Section 2.7, including Figure 2-22(a to c);

(f) Adequate strength and stiffness of lateral bracing/restraining elements; and

(g) Adequate consideration of global frame stability.

With regard to Item 3.1(c), (d), and (f), the need to adequately check conditions in the structure cannot be overemphasized, particularly when computer software is utilized. Proper representation of stability in the computer model may not occur.

For example, the computer software may consider restraint at node points, which may not occur or may not be sufficiently stiff to prevent instability. Therefore, it is incumbent on the structural engineer to thoroughly check all potential instability conditions.

3.2 WIND

Considerations relating to wind forces, in addition to those listed for gravity, include (1) accuracy of determining the wind forces including orthogonal effects, and (2) wind forces inducing lateral forces, causing additional secondary stresses; one example is shown in Figure 3-1(a), where eccentric connections for wind girts could cause torsional buckling issues if not adequately considered in the design. Figure 3-1(b) indicates a section of a bridge where wind forces can cause lateral sway, which may cause instability to the top chord in compression again if not adequately considered in the design.

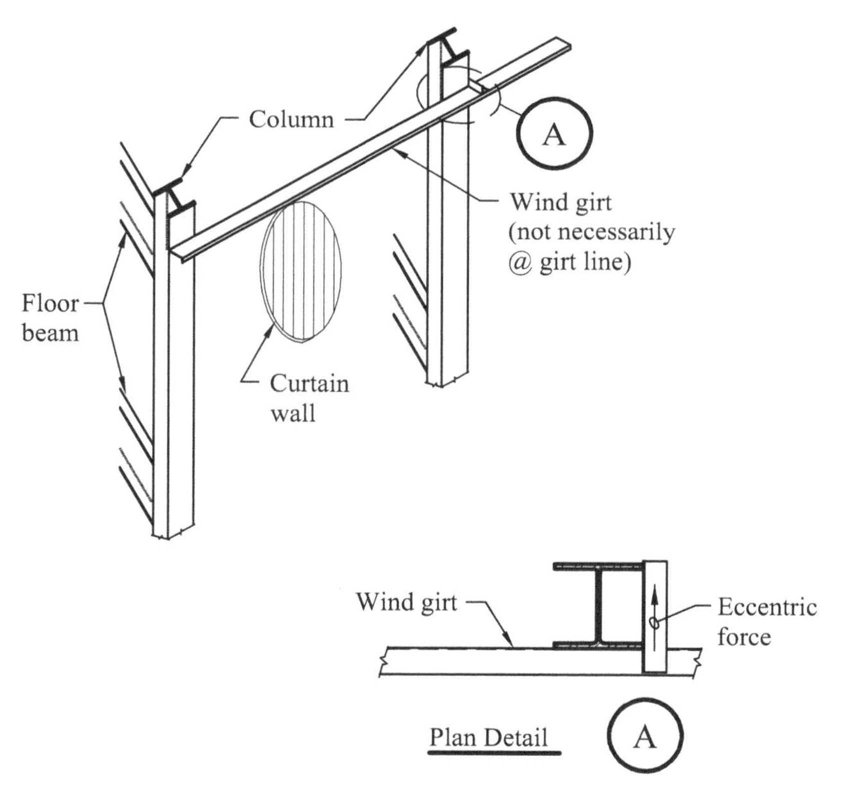

Figure 3-1.(a) Eccentric connection of wind girt: 3D view at face of structure.

Section 2.9, indicates the substantial local buckling and global buckling at the culmination of a girder-to-column moment frame connection test, utilizing a bolted flange plate moment connection. This test satisfied the testing requirements given in Appendix S of ANSI/AISC 341 Seismic (AISC 2005b). In addition to causing member instabilities, large cyclic deformations and local buckling can result in low-cycle fatigue.

An examination of the results of many beam-to-column tests carried out revealed that the flange and web local buckling has actually helped to achieve the inelastic rotation required by Appendix S of ANSI/AISC 341 Seismic (AISC 2005b). This is because the resulting distortions that occur, as a result of these phenomena, tend to increase the inelastic rotation of the girder to the column test specimen.

2. *Global stability (lateral torsional buckling) of the portions of a girder subject to elastic behavior*: Historical background, on addressing girder stability, is given in Chapter 2, Section 2.6. For a moment beam in seismic areas, yielding occurs in close proximity to the girder-to-column connection. Beyond the yield zone, the girder is in the elastic range. However, it should be recognized that the girder, while behaving elastically in this region, is affording restraint to the portion of the girder within the yield zone. As yielding progresses, flange local buckling develops, incurring lateral displacement. Demand on the portions of the girder, subject to elastic behavior, increases. On the basis of the principles of mechanics and observations of tests, a minimum of two braces/lateral restraints to the bottom flange should be provided.
3. *Effects of welding*: As stated in Chapter 4, significant distortion that occurs because of weld shrinkage can trigger the onset of buckling.
4. *Panel zone yielding*: Yielding of the panel zone has had beneficial results on the performance of some connections in satisfying the requirements of Appendix S of ANSI/AISC 358 Seismic (AISC 2005a, 2016a) by making a significant contribution to the rotational capacity. However, panel zone yielding may not be predictable because the yield strength of steel, for ASTM Grade 50, can vary from 50 to 65 ksi (344.7 to 448.1 Mpa). When applied to both the girder and the column, the girder/column strength ratios can theoretically vary from 0.76 to 1.3. If panel zone yielding occurs, the ability of the column to provide torsional restraint is reduced. This increases the tendency of the column to twist in an uncontrolled manner. Column twisting (rotation in the horizontal plane) further enhances the potential for buckling of the girder. Buckling of the panel zone is possible because of the high shear that can occur. This is partially accommodated by tension field action dependent on the strength and stiffness of the column flanges and continuity plates, if used.
5. *Column yielding*: ANSI/AISC 341 Seismic (AISC 2005b, 2016b) provisions require the columns to have greater strength than the girders. This provision is intended to minimize the potential for column yielding. However, the phenomenon known as *column moment magnification*, demonstrated by

time history analysis for seismic events, indicates that column moments can be significantly enhanced in multistory buildings because of the contributions of higher modes. Thus, the possibility of column yielding may be significant unless column/girder strength ratios are greater than 2.0. Column yielding raises two of the following concerns: (1) Columns cannot be readily braced near the hinge area. (2) No column-to-girder tests were carried out where the column yielded. Thus, there is a concern how steel moment connections will perform in this event.

6. *Lateral restraint/bracing details*: As stated in Chapter 2, Section 2.4, strength and stiffness is required to provide restraint of the compression flange. According to the principles of mechanics, any out-of-plane distortion will cause out-of-plane forces that have to be resisted. Therefore, sufficient strength needs to be provided. Any distortion that continues to magnify without control will cause instability. Thus, sufficient stiffness needs to be provided. Recommendations pertaining to lateral bracing at moment frame girders include the following:
   (a) Minimum of two braces/lateral restraints needs to be provided.
   (b) Lateral bracing need not necessarily be immediately adjacent to the plastic hinge (see Section 4).
   (c) Brace to the girder should be perpendicular to the girder to prevent the brace from being subjected to an additional force because of the component of the girder flange compressive force.
   (d) Use double panel zone plates, combined with continuity plates, according to ANSI/AISC 341 Seismic (AISC 2005b), Figure C-I-9.3(c), to provide significant torsional rigidity, thus reducing column twisting. ANSI/AISC 341 Seismic (AISC 2016b), Figure C-E3.5, shows a similar detail.
   (e) Column/girder strength ratios should be significantly greater than 1.25 specified for multistory buildings to reduce the potential for column yielding because of column moment magnification.
   (f) Measures need to be provided to control column twisting, beam flange, and web local buckling, particularly deep columns.
7. *Controlled local buckling*: Control of local buckling can significantly benefit the ductility of connections and minimize their potential to fracture. Control of web buckling can be addressed by adding horizontal stiffeners at the centerline of the web that is often used in bridge design. Control of flange buckling with clamped plates was carried out on one test on a girder-to-column moment connection using a bolted flange plate with elongated holes in the beam flanges [Figure 3-4(a)]. Although this has not been formally published, the writer supervised the testing and found that a control of buckling of girder flanges, utilizing clamp plates, can result in good performance [Figure 3-4(b)].

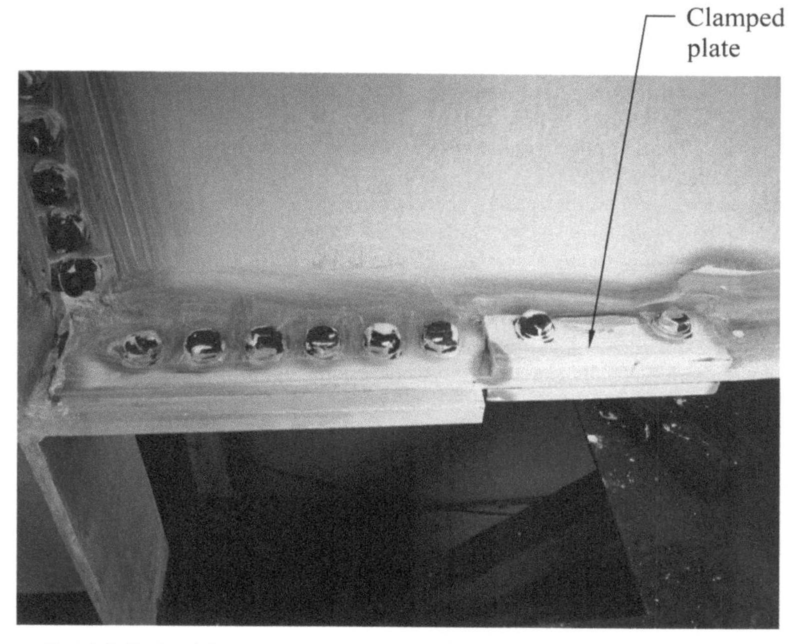

Figure 3-4.(a) Bolted flange connection with clamp plates to control buckling: Before testing.

Figure 3-4.(b) Bolted flange plate connection with clamp plates to control flange: After testing.

References

AISC (American Institute of Steel Construction). 2005a. *Prequalified connections for special and intermediate steel moment frames for seismic applications.* ANSI/AISC 358 Seismic. Chicago: AISC.

AISC. 2005b. *Seismic design provisions for structural steel buildings.* ANSI/AISC 341 Seismic. Chicago: AISC.

AISC. 2016a. *Prequalified connections for special and intermediate steel moment frames for seismic applications.* ANSI/AISC 358 Seismic. Chicago: AISC.

AISC. 2016b. *Seismic design provisions for structural steel buildings.* ANSI/AISC 341 Seismic. Chicago: AISC.

Bondy, K. D. 1996, "A more rational approach to capacity design of seismic moment frame columns." Oakland, CA: Earthquake Engineering Research Institute.

Maranian, P., and A. Dhalwala. 2019. "Considerations regarding the repair & retrofit of existing welded moment frame buildings." In *Proc., Structural Engineers Association of California Convention 2019, Squaw Creek, California.* Sacramento, CA: SEAOC, 396–409.

Maranian, P., R. Kern, and A. Dhalwala. 2012. "Considerations on buckling and lateral bracing issues with an emphasis on steel moment frames in seismic areas." In *Proc., 6th Congress on Forensic Engineering, San Francisco.* Reston, VA: ASCE, 1–14.

Paulay, T., and M. J. N. Preistley. 1992. *Seismic design of reinforces concrete and masonry buildings.* New York: Wiley.

CHAPTER 4

Effects of Steel Production, Fabrication, Details, and Others

4.1 STEEL PRODUCTION

The steel-making process can have significant influences on steel properties across the section of the steel member. During the rolling process, plastic deformation takes place such that the material is work-hardened. The temperature at which rolling is applied can significantly affect the steel properties. Higher temperatures [about 1,830 °F (998 °C)] produce coarse-grain steel, which tends to be less ductile and more brittle. Rolling at lower range temperatures [1,650 °F to 1,830 °F (898 °C to 998 °C)] will result in fine grain steel that tends to provide improved ductility. Rapid cooling can result in a steel structure known as *martensite*. Martensite has high strength and is very hard but brittle, with low ductility.

The steel-making process considers the chemistry, rolling practice, and thermal changes to control the desired steel properties. Nevertheless, an appreciable variation of properties can occur. For example, with regard to wide-flange members, the web has a 4% to 7% higher yield strength than the flanges because of the web having a finer grain structure and higher carbon content. Furthermore, the uneven cooling of structural steel sections during the rolling process can result in significant residual stresses. The magnitude of the residual stresses is dependent on temperature during rolling, cooling rate straightening, and the steel chemistry. Figure 4-1 (Johnston 1976, p. 55) indicates residual stress distribution in rolled wide-flange shapes. The outer region of the flanges tends to be in compression with the remaining inboard region of the flanges, along with all the web, being in tension. Compression at the outer edges may contribute to a reduction in both global stability and local buckling. The effects of residual stress, along with initial imperfections, have been much researched, commencing in the 1940s both in the United States and in Europe. Residual stresses have been found to have a significant effect on column strengths, particularly for wide-flange members. Research carried out at the University of Michigan (circa 1967), utilizing computer analysis, which included the effects caused by both residual stresses

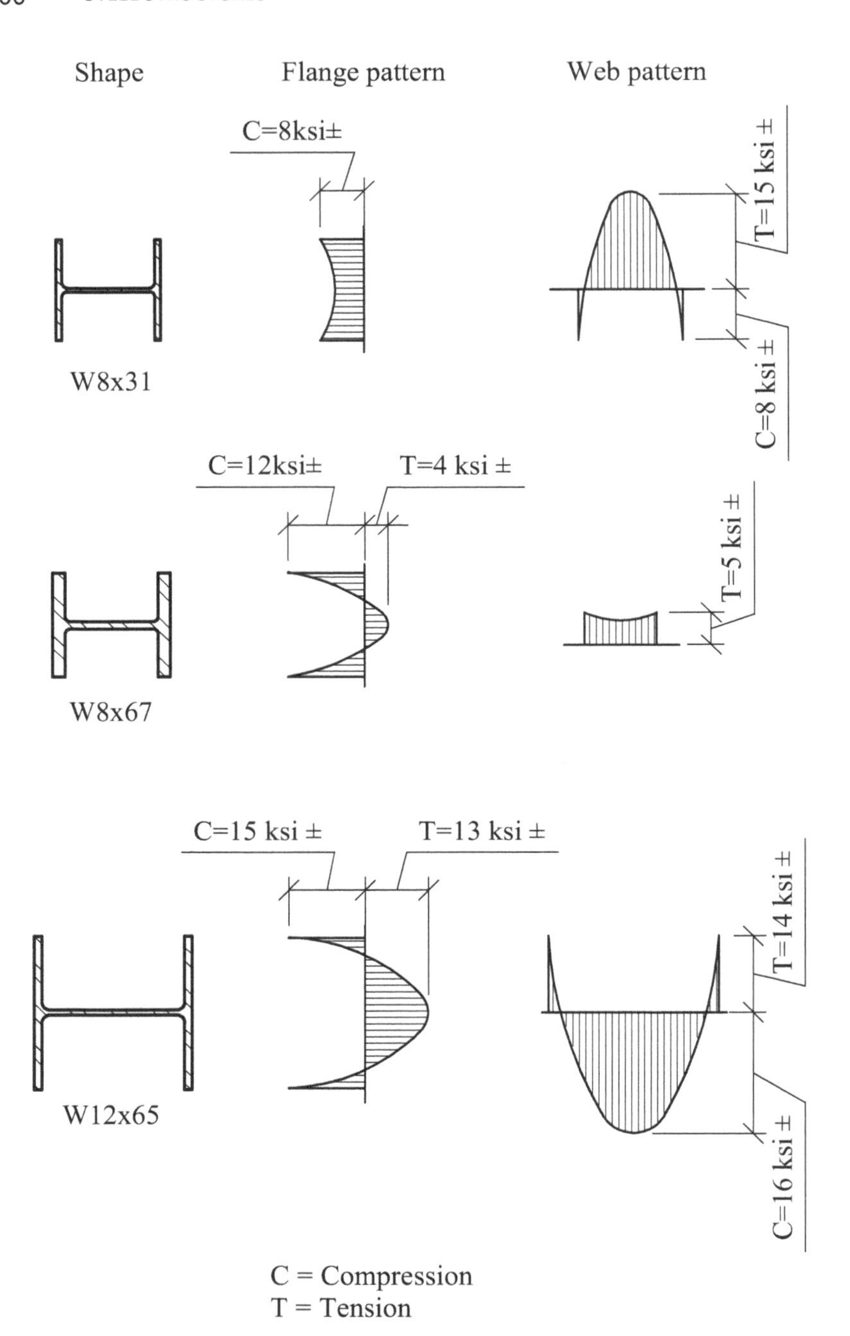

Figure 4-1. Residual-stresses distribution in rolled wide-flange shapes.

Source: Reproduced from Johnston (1976, p. 55), courtesy of Wiley.

Note: Also refer to Figure 2.8 (Johnson 1976, p. 55).

and initial curvature, showing a greater decrease in strength with increasing size and weight (Johnston 1976, pp. 53–56). This research led to the development of a column formula, as discussed in Chapter 2, Section 2.2.

Cold straightening of columns, carried out either by the rotary straightening process or by the gag straightening process to meet camber and sweep tolerances, redistributes the residual stresses caused during the thermal changes during rolling. Rotary straightening (also known as *rotorizing*) is used on lighter members (150 lb/ft or less) (223.2 kg/m or less), where a train of rollers apply bending, back and forth, to straighten the member, and can significantly reduce the residual stresses. The gag straightening process, used on heavier members, involves forces applied at more concentrated areas at the edges of the flanges.

Stresses because of gag straightening can be close to yielding, such that they can reverse the residual stresses at the flange edges from compression to tension.

Thus, the effects of cold straightening, by modifying residual stresses, can be such that they increase the column capacity with the roller straightening process, providing a greater increase than that with the gag straightening process (Johnston 1976, p. 62).

4.2 STEEL MATERIAL PROPERTIES

The stress–strain properties of steel can have an appreciable effect on buckling performance. The modulus of elasticity, representing the stress–strain relationship in the elastic range, is relatively constant for most steels at about 29,000 ksi (199,947 Mpa). However, yield strengths are affected by the strain rate such that increases in yield stresses can be of the order of 4 to 5 ksi (27.5 to 34.4 Mpa) at loading rates of 100 ksi/min (689.4 Mpa/min). Therefore, buckling performance can be affected by short-term increases in yield stress because of the rate of loading such that the slenderness ratio (l/r), at which Euler buckling occurs, is reduced.

Yield strengths, which are, in general, measured at 0.2% strain, are also affected by temperature such that yield strengths reduce with increasing temperature. Thus, consequentially, buckling performance can be affected by decreases in yield stress such that the slenderness ratio (l/r), at which Euler buckling occurs, is increased.

Beyond yield, increase in stress with an increase in strain is called strain hardening. This is represented as the strain hardening modulus and is also termed the *tangent modulus*, as discussed in Chapter 2. Again, as discussed in Chapter 2, variations in the strain hardening modulus (tangent modulus) can have a significant influence on the behavior of short columns.

Steel properties, including yield strength, ductility, and fracture toughness, are affected by the metallurgy of steel. An increase in carbon increases the strength and the hardness of steel, while decreasing ductility, fracture toughness, and weldability. Carbon does have a tendency to segregate, and a higher carbon content has been found in the web of wide-flange members, providing a higher yield strength than that with the flanges. Manganese also strengthens steel and improves fracture toughness. Phosphorus, which has a strong tendency to segregate, also increases the strength of steel but decreases ductility and toughness. Nickel improves strength and fracture toughness.

Heat treatment of steel, carried out subsequent to rolling, also affects the properties of steel. Annealing, typically heating at above 670 °C (1,240 °F) for many hours with slow cooling, reduces strength, while improving fracture toughness. Normalizing is similar to annealing, except that the steel is heated to 870 °C (1,600 °F) and the steel is cooled more rapidly in still air. This process results in greater strength and improved fracture toughness. Quenching involves heating the steel to above 730 °C (1,330 °F) and quenching in oil or water. This results in hardened steel that can be very brittle. Subsequent heating and then slow cooling, known as *tempering*, reduces the brittleness of steel such that the combination of strength and improved fracture toughness can be attained.

Thus, with regard to postyield buckling behavior, performance is affected by the postyield stress–strain curve, sometimes known as the *backbone* curve (Figure 4-2). This primarily affects the local buckling of flanges and the web. A flatter backbone curve (smaller strain hardening modulus), usually more prevalent with mild steel, will cause greater postyield buckling than a *roundhouse* backbone curve occurring with high-strength steels.

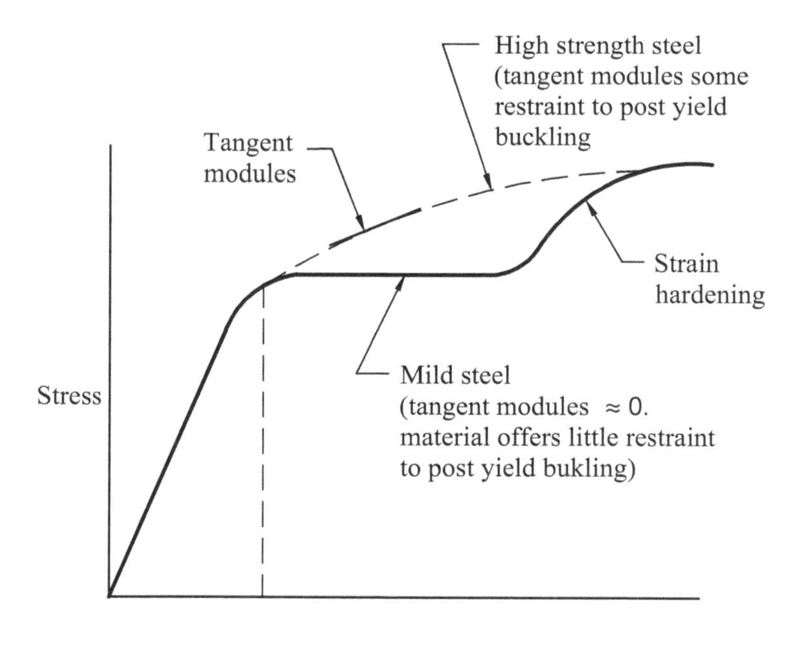

Figure 4-2. Stress–strain diagrams—effects on postyield buckling.

Figure 2-31(a, b) discussed in Chapter 2, Section 2.8, indicate local buckling of flanges and web during beam to column tests on bolted flange plate (BFP) moment connections. A cyclic load test, simulating a seismic event, defined in Appendix S of ANSI/AISC 341 Seismic (2005), was applied to the specimen, eventually yielding the beam at a high in-plane rotation. The local buckling is indicative of postelastic local buckling. Similar performance has been noted with some other connections, including the RBS connection.

As can be seen, significant local buckling occurred in a wavelike manner. The test satisfied the requirements in Appendix S of ANSI/AISC 341 Seismic (2005) for steel moment frame connections, with failure eventually occurring as a result of a fracture in one flange because of local cycle fatigue accentuated by the local buckling.

4.3 WELDING

It has become well-established that welding, such as for plate girders and box sections, can cause high residual stresses at or near the welds.

Residual stresses can be as high as the yield stress or greater. The residual stresses, because of welding, modify the residual stresses arising from the mill rolling process, where the milled-rolled edges have compressive stresses. Welding increases the compressive stresses at the edges, thus decreasing the column strength. However, with flame (oxygen) cut plates, tensile stresses occur at the edges such that, although welding causes compression at the edges, the resulting stresses at the edges may remain in tension.

This is illustrated in Figure 4-3 (derived from Johnston 1976, p. 59). Welding has more adverse effects on small and medium-sized members in comparison with heavy members (Johnston 1976, pp. 57–60).

The process of welding can also result in distortion, which, in turn, can cause buckling or initiate buckling to occur at lower than expected forces. Distortion and buckling distortion are illustrated in Figure 4-4 (derived from Masabuchi 1980, p. 236). Shrinkage occurs during the thermal cycle during welding, which causes both transverse and longitudinal dimensional changes to the line of the weld and angular distortion. The amount of shrinkage is dependent on factors such as the welding process, geometry sequence of heat input welding, and so on.

Weld distortion can also cause bending, which, particularly for thin plates, can reach a critical stage such that buckling occurs. Buckling distortion can have more than one deformed mode shape, and the amount of distortion can be much more than bending distortion.

An excellent discussion on this is given in Masabuchi (1980). Masabuchi discussed tests carried out by Watanabe and Sato on two plates, 40 in. (1 m) by 80 in. (2 m) less than 0.16 in. (4.5 mm). Residual stresses are built up, as shown in Figure 4-5 (derived from Masabuchi 1980, p. 301), such that there are significant tension stresses at, and immediately adjacent to, the weld but with significant

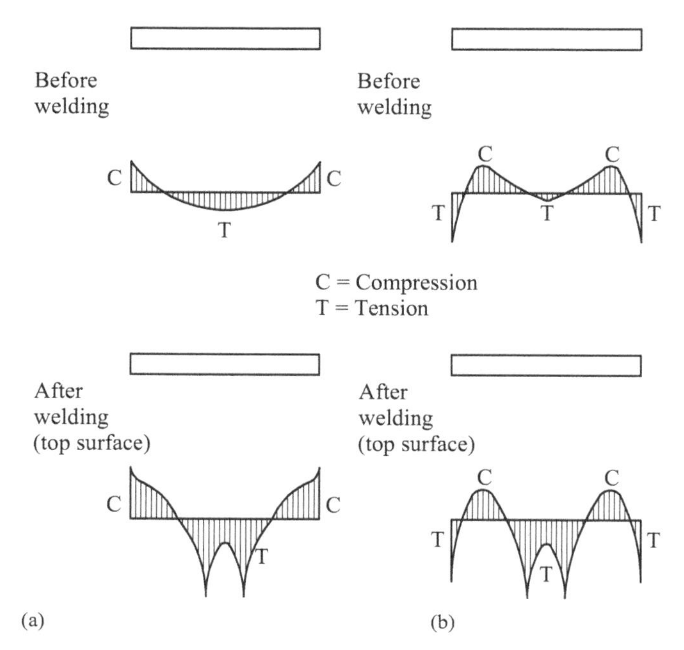

Figure 4-3. Qualitative comparison of residual stress in as-received and center-welded universal mill and oxygen-cut plates: (a) universal mill plate, (b) oxygen-cut plate.

Source: Reproduced from Johnston (1976, p. 59), courtesy of Wiley.

compression stresses occurring toward the edges of the plate. Watanabe and Sato derived curves for the critical thickness versus the plate width, B, for different aspect ratios, B/L. It can be seen that the effects of distortion because of welding not only significantly reduce the buckling strength of components but also reduce the global buckling capacity of the builtup members.

Figure 4-6 shows the residual stress distribution that can occur in the "H" and box sections because of welding. The extreme outer regions of the flanges of the "H" section, along with the center region of the web, have compressive residual stresses; the remaining regions, at and near the welded connections, are in tension. These conditions tend to cause distortion, which reduces the global buckling capacity of the hinged steel column, as shown in Figure 4-7 (derived from Figure 8.6 in Timoshenko 1955).

Figure 4-8 (derived from Maranian 2010), shows the amount of distortion that occurs from welding two plates together. As much as 3 in. distortion occurred in 12 in.

Factors that can assist in reducing distortion are as follows:

(a) Increasing plate thickness,

(b) Reducing unsupported spans,

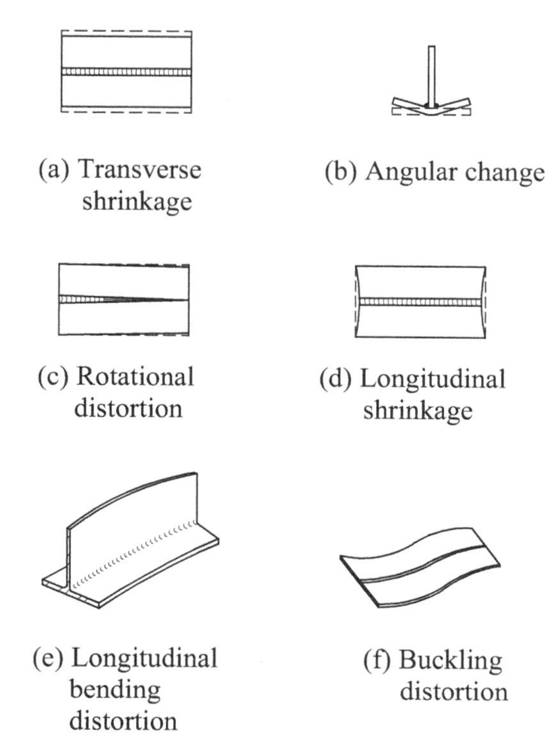

Figure 4-4. Various types of weld distortion: (a) transverse shrinkage, (b) angular change, (c) rotational distortion, (d) longitudinal shrinkage, (e) longitudinal bending distortion, (f) buckling distortion.

Source: Derived from Masabuchi (1980, p. 236).

(c) Reducing weld size, and

(d) Reducing heat input.

Masabuchi (1980) discussed the importance of selecting the appropriate structural and welding properties to control distortion.

A consideration of these factors is illustrated in Figure 4-9 (Masabuchi 1980, p. 625) relating heat input to plate thickness.

With reference to Figure 4-9, for a given thickness, the first weld pass will cause a distortion of $\partial 1$. The second weld pass will cause a distortion of $\partial 2$; the distortion is increased but still remains small and acceptable. The third pass may cause a considerable distortion of $\partial 3$, which would not be acceptable. A thicker plate will reduce the distortion but will involve an increased weld thickness.

Masabuchi (1980, pp. 312–325) discussed several methods for distortion-reduction in weldments. Primarily, along with a careful selection of the joint

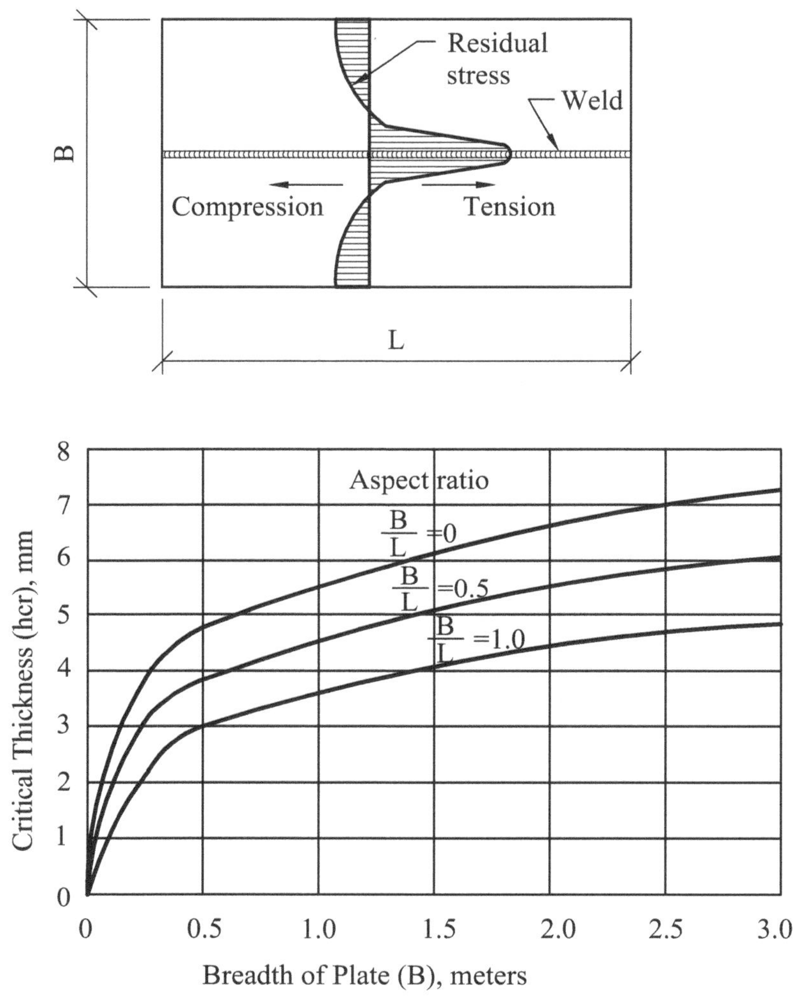

Figure 4-5. Critical thickness for buckling distortion of butt weld.
Source: Derived from Masabuchi (1980, p. 301).

details, the weld and heat input need to be kept to a minimum. Adopting weld joint details such as double-bevel welds in lieu of a single bevel helps reduce weldments and can also appreciably help control distortion. Also, by sequencing weld deposition, balancing between one bevel and the other, can enable the control of the distortion. Larger-diameter weld electrodes can help reduce distortion. Other methods include applying external restraints, use of thermal-pattern control, stretching and heating, and differential heating.

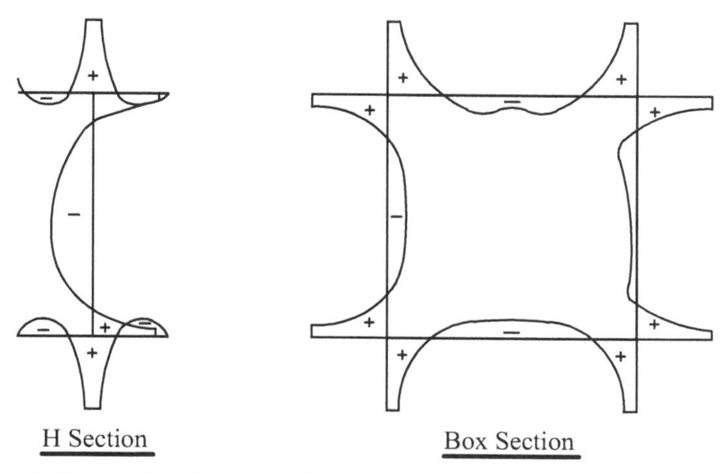

Figure 4-6. Typical distributions of longitudinal residual stress in an H-section and a box section fabricated by welding.

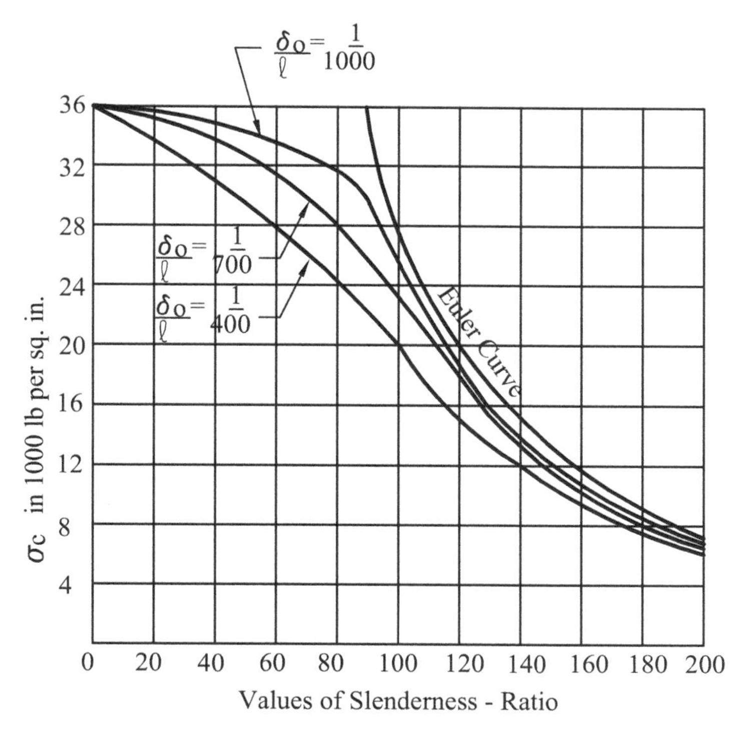

Figure 4-7. Effect of initial distortion, on the buckling stress of a hinged steel column under compressive loading.

Source: Derived from Figure 245 in Timoshenko (1955, p. 276).

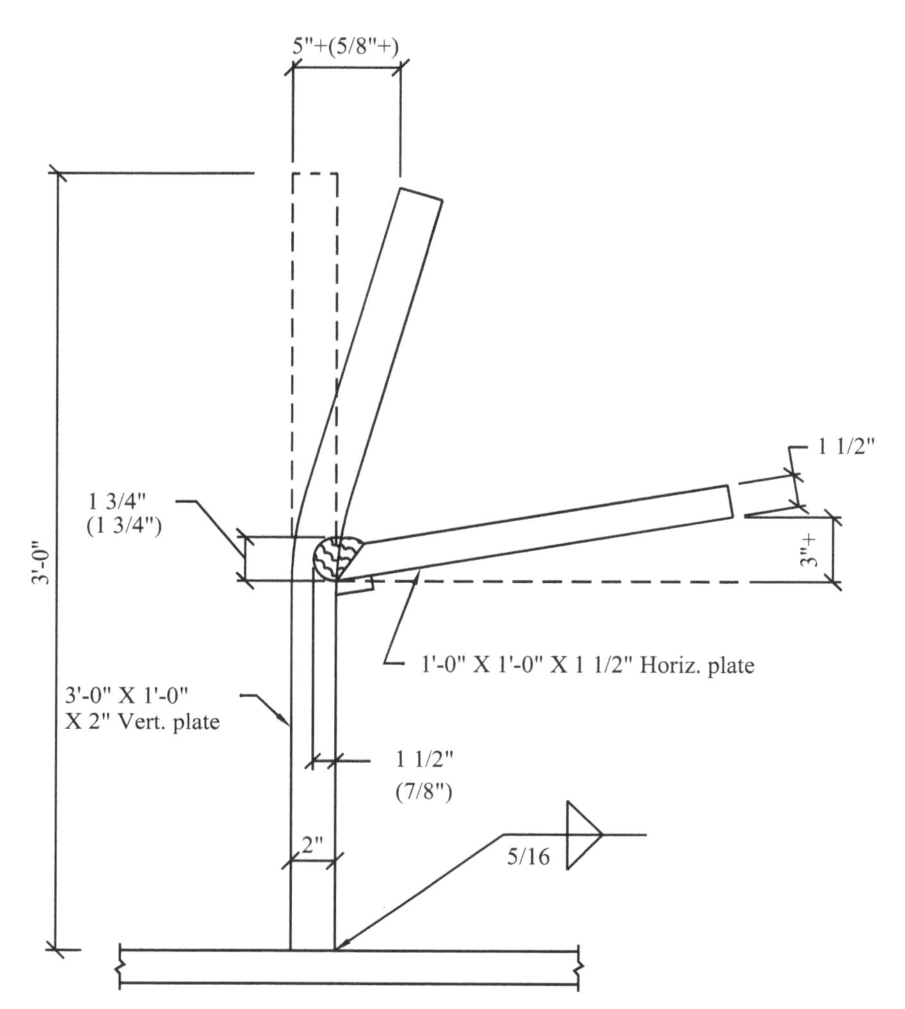

Figure 4-8. Distortion of weldment due to unrestrained shrinkage.
Source: Reproduced from Brandow and Maranian (2001), courtesy of ASCE.

4.4 DETAILS WITH UNINTENDED ECCENTRIC CONDITIONS

Connection details, including beam-to-column connections, splice details, braced frame details, and so on, can cause unintended eccentricities that can reduce the buckling and lateral torsional buckling strength of the members.

Examples of details, where inadvertent eccentricities that may cause instability can occur, follow.

Figure 4-10 illustrates a welded flange plate moment connection. As can be seen, the offset between the flange plates and the center of the beam flanges, because of tolerances in rolling and fit-up, results in eccentric forces' being

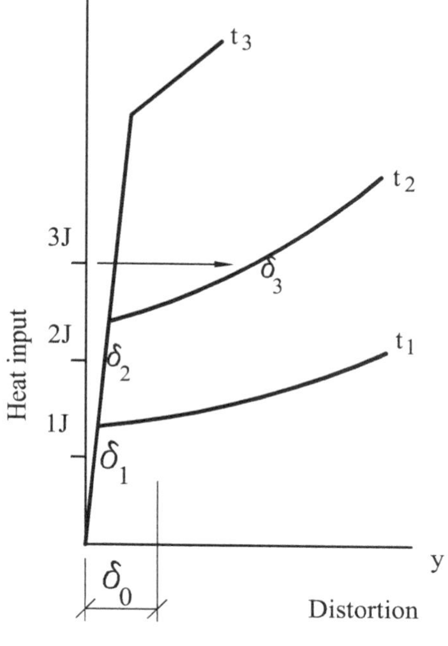

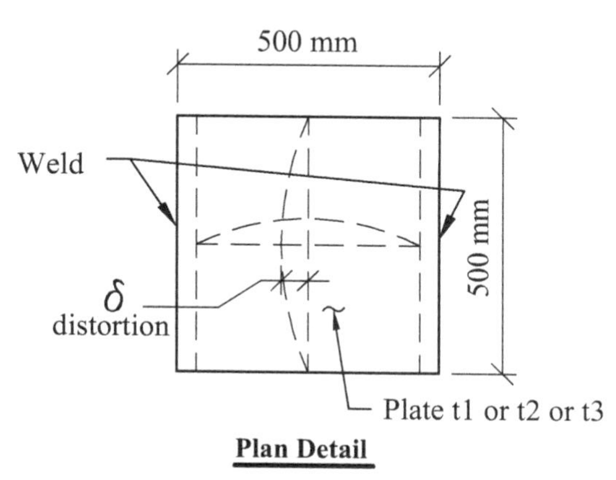

Figure 4-9. Effects of heat input and plate thickness on out-of-plane distortion due to buckling.

Source: Derived from Masabuchi (1980, pp. 302 and 625).

applied. This eccentricity enhances the potential for local buckling of the beam flanges. An improved detail using thick continuity plates is shown.

Figure 4-11 indicates a girder splice detail with single flange plates. Again, eccentric forces occur, which can enhance the potential for local buckling of the beam flange. A better detail using plates each side, minimizing eccentricity, is shown.

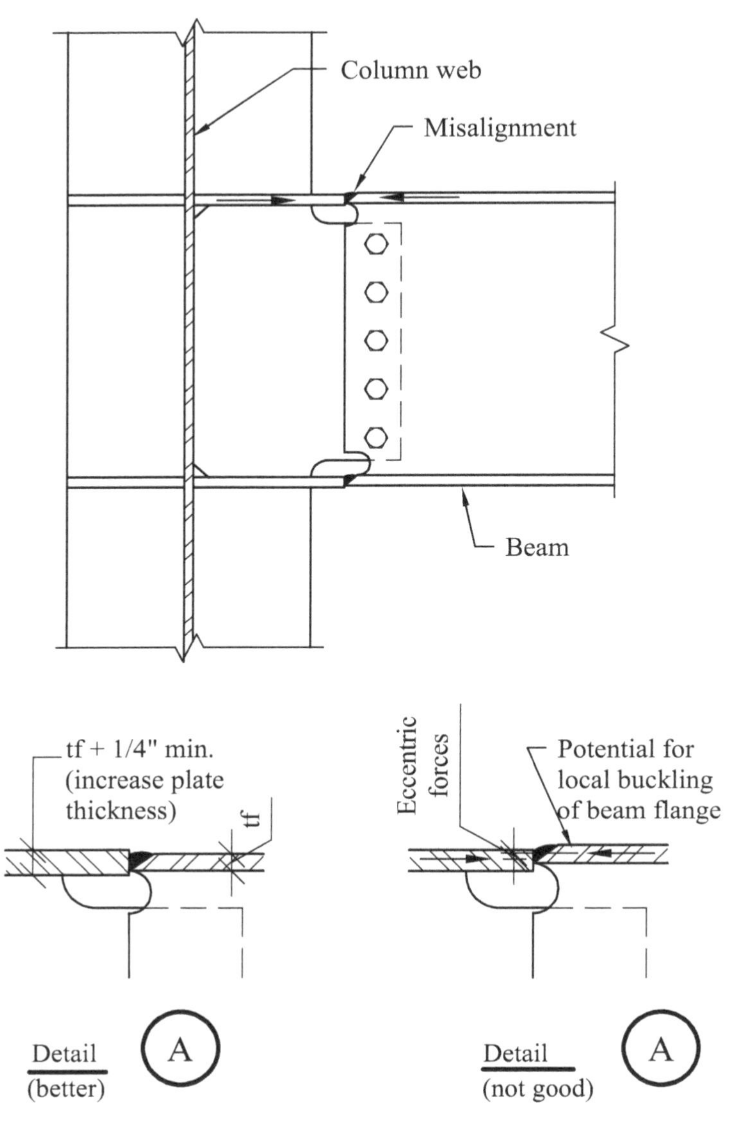

Figure 4-10. Section—misaligned beam-to-column connection.

Figure 4-12(a) indicates a gusset plate detail for a brace member with a lack of thought in the detail allowing eccentricity to occur. The eccentricity induces bending in the brace, thus reducing the buckling capacity of the brace. The gusset plate, because of the eccentricity of the load, is also subject to larger stresses. A better detail is shown in Figure 4-12(b).

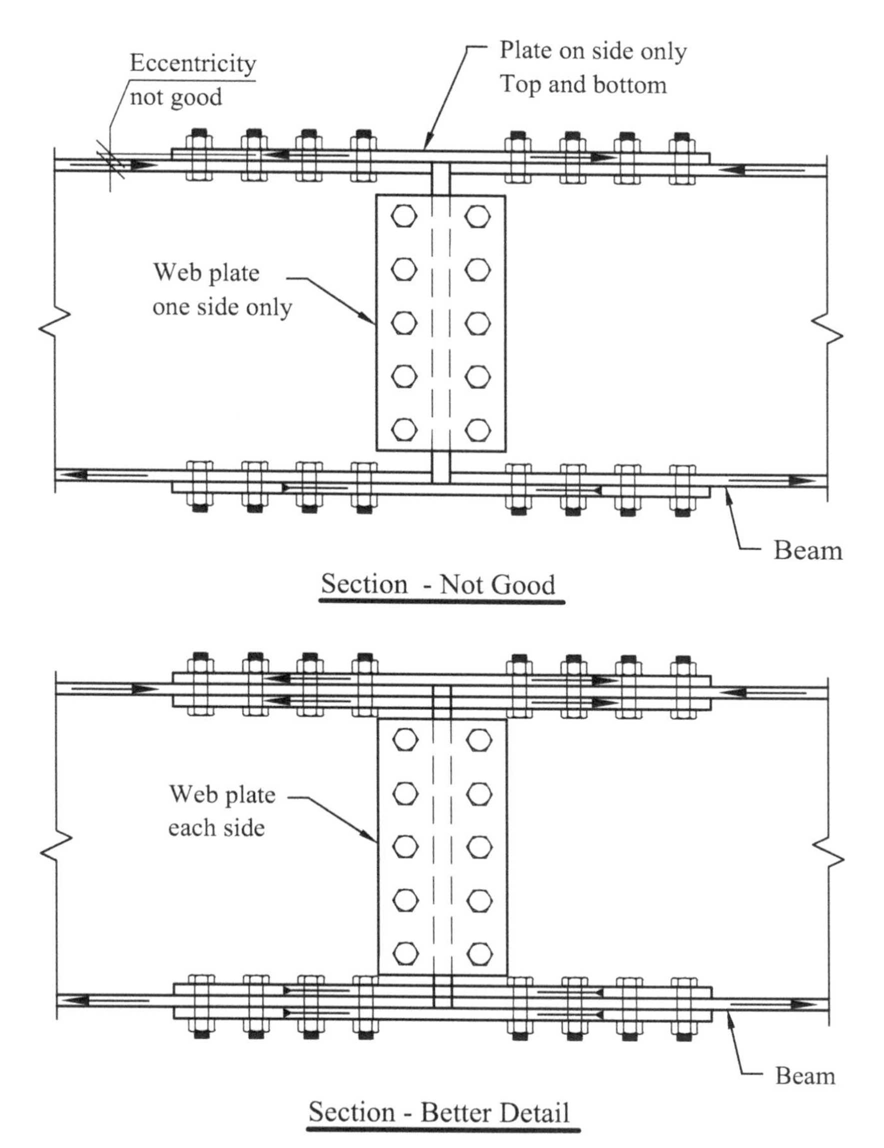

Figure 4-11. Section—offset at girder splice connections.

Figure 4-13 indicates a misaligned stiffener plate condition. The compression load has to be transferred through bending of the flange, which induces secondary bending in the flanges and web stiffener plates, thus reducing its buckling capacity.

Figure 4-14 indicates a shore not correctly located over the centerline of a beam. The resulting eccentric load may cause instability.

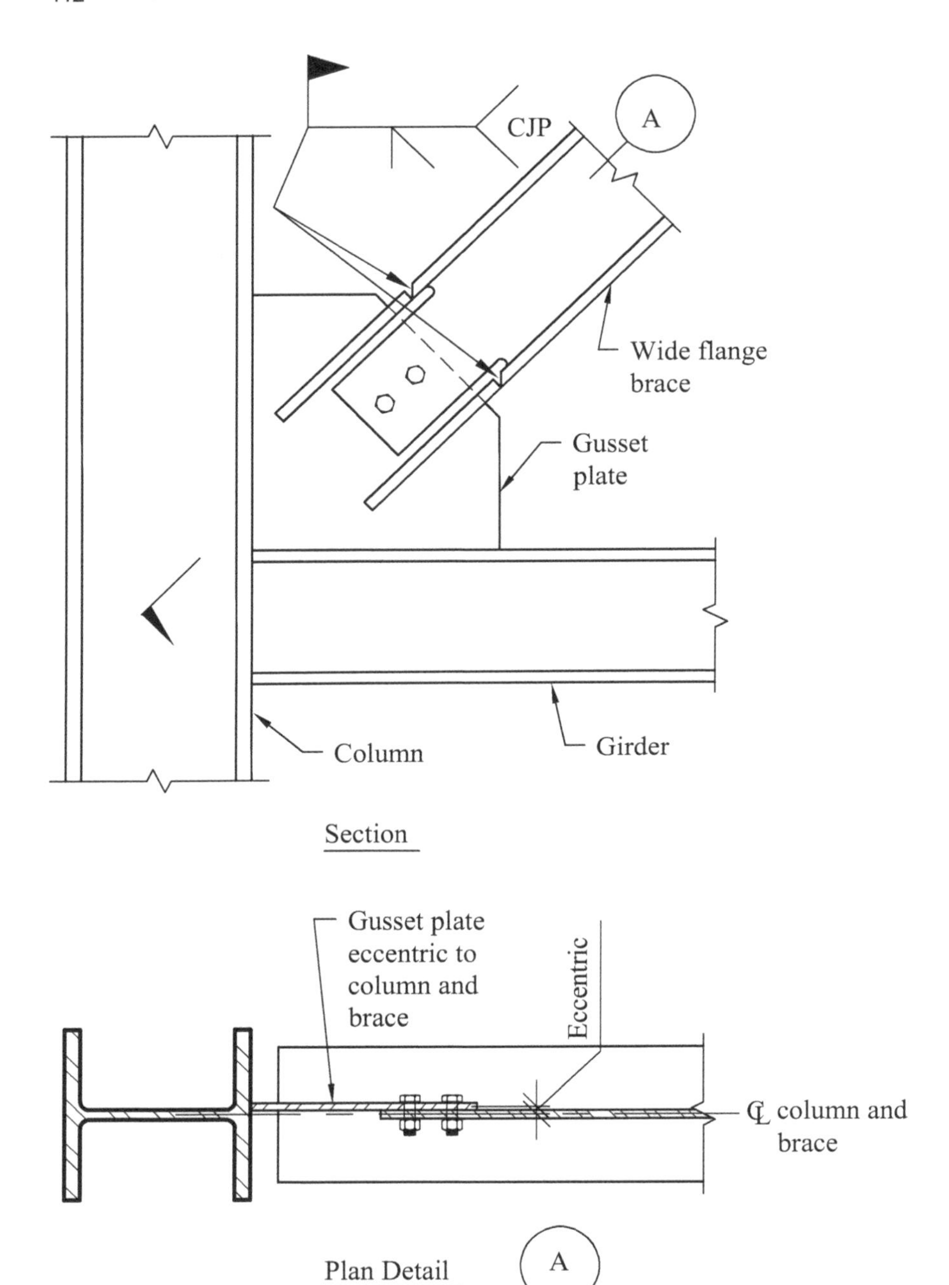

Figure 4-12.(a) Brace connection–Gusset plate eccentric.

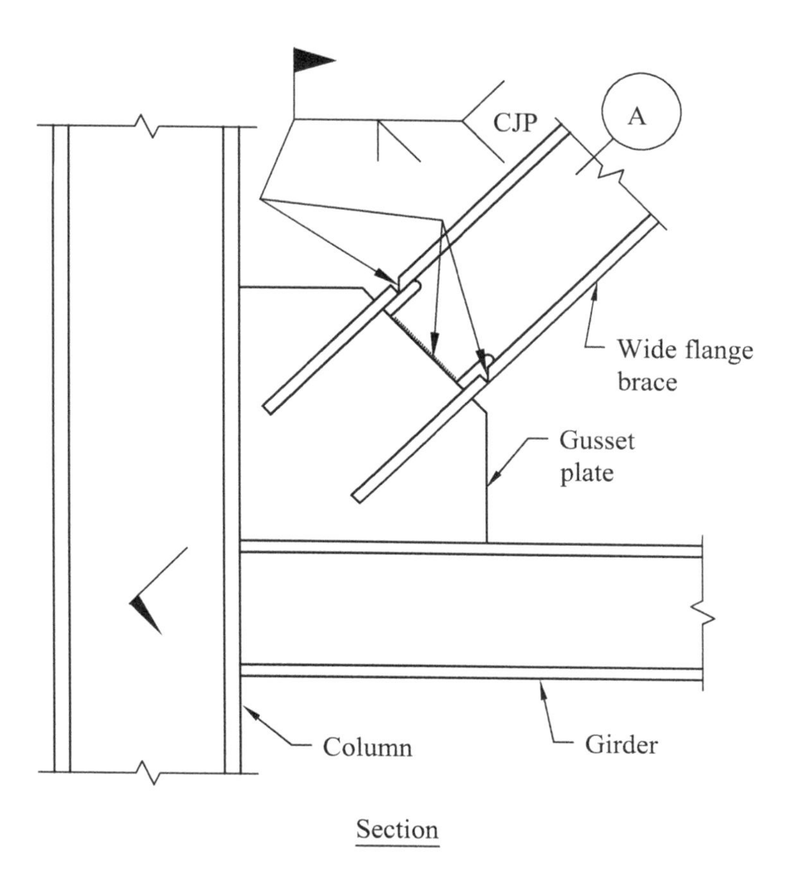

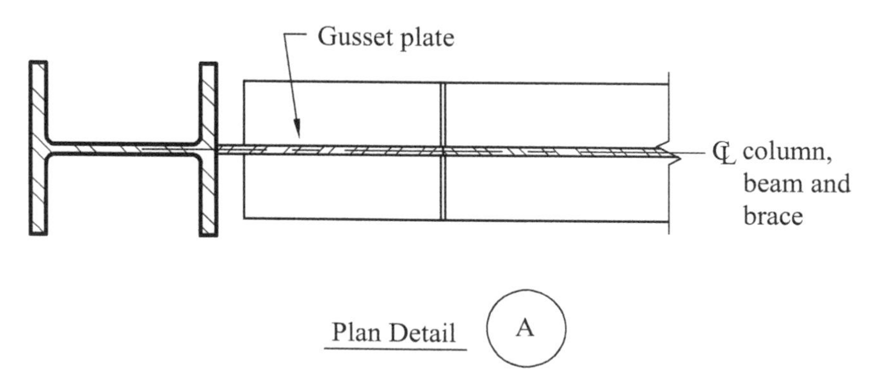

Figure 4-12.(b) Brace connection–concentric condition.

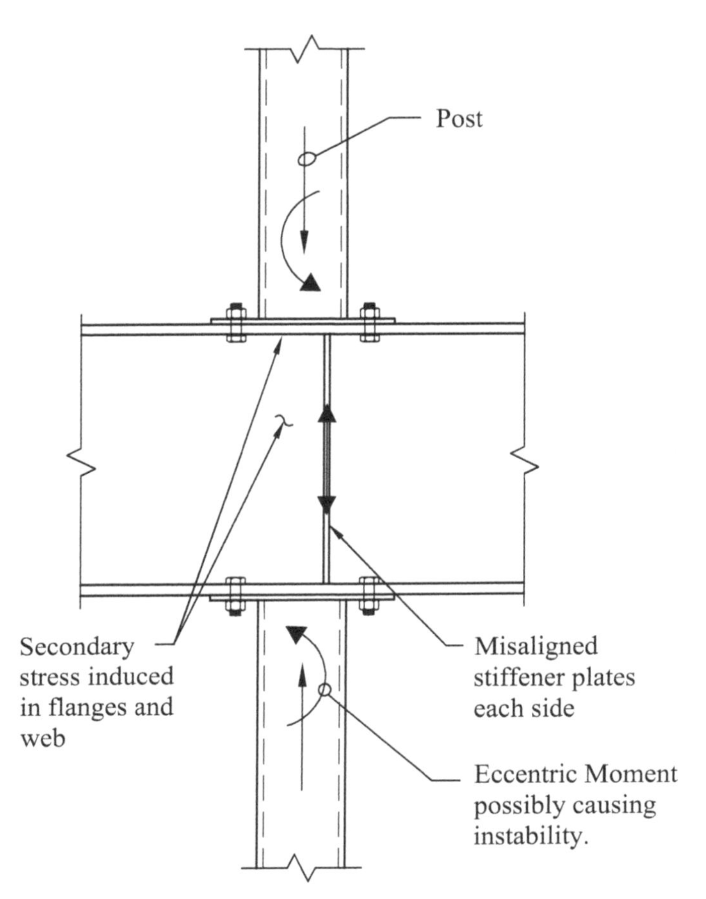

Figure 4-13. Misaligned stiffeners.

4.5 BEAM COPES: POTENTIAL TO CAUSE LOCAL BUCKLING

Beam copes are often not fully identified in the design process but established during the shop-drawing process.

The potential for shear buckling is shown in Figure 4-15(a). Copes in beams can become large because of factors that include the following:

(a) Girder flanges; being wide [Figure 4-15(b)],

(b) Beam and girder depths' being the same or similar such that copes are required at both the top and the bottom of the beam [Figure 4-15(c)],

(c) Skewed beam to girder connection [Figure 4-15(d)], and

(d) Sloping beam to girder connection such as may occur at roofs [Figure 4-15(e)].

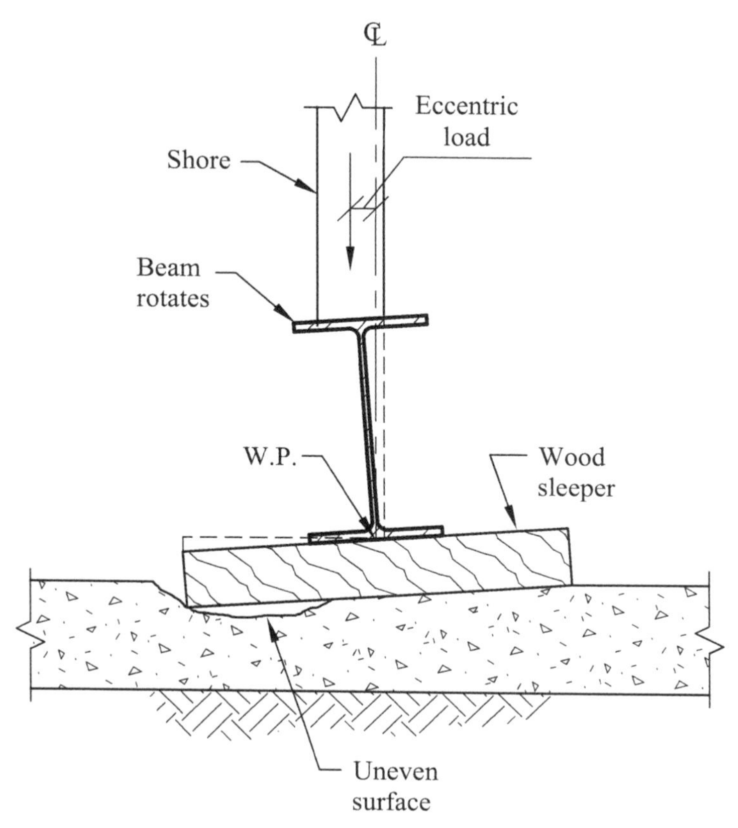

Figure 4-14. Section—unintended eccentricity causing instability.

An excellent discussion is given by Dowswell (2018) with regard to the changes made in the *Steel Construction Manual* (AISC 2017).

Where copes are short and/or have slender webs, they tend to fail by shear buckling, typically at about 45 degrees, as shown in Figure 4-15(a), before shear yielding occurs. However, when copes become large, localized secondary bending stresses can occur because of a loss of one or both flanges [Figure 4-15(b, c)]. The web is no longer restrained such that, where copes are long, flexural local buckling, in combination with shear stresses, can occur prior to shear yielding. Connections involving the beam at an acute angle to the girder can significantly enhance the size of the cope, as shown in Figure 4-15(d). Also, sloping roof conditions can further increase the size of top flange copes, as shown in Figure 4-15(e). According to Dowswell (2018), most instabilities with large copes are caused by a combination of both shear buckling and flexural local buckling. Instabilities can be increased further if axial compressive forces occur.

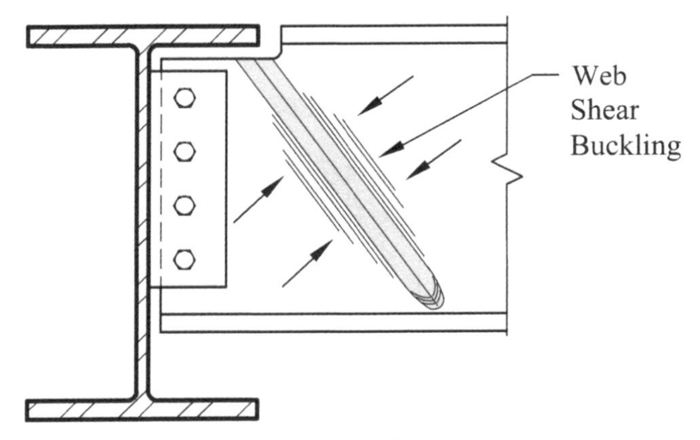

Figure 4-15.(a) Beam copes, shear buckling: Section.

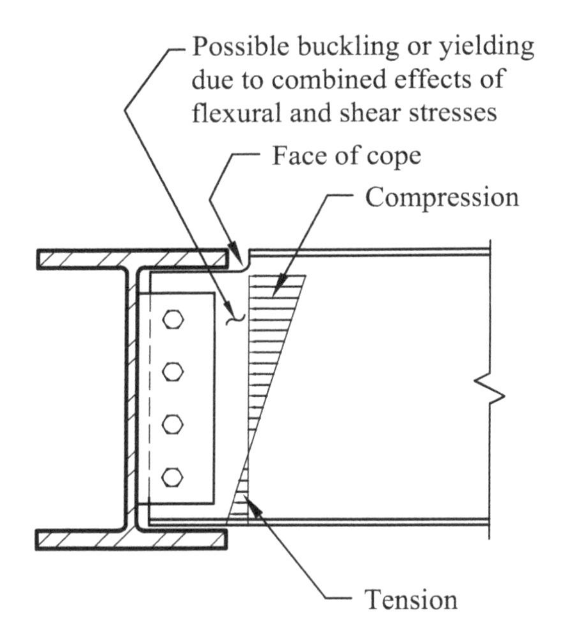

Figure 4-15.(b) Large cope due to girder flange being wide: Section.
Source: Reproduced from Dowswell (2018), courtesy of AISC.

Another type of failure involving a combination of block shear and block shear buckling occurs where shallow depth shear plate connections are provided; this is also discussed in Dowswell (2018) [Figure 4-15(f)]. Design guidelines on these issues are provided in the *Steel Construction Manual* (AISC 2017).

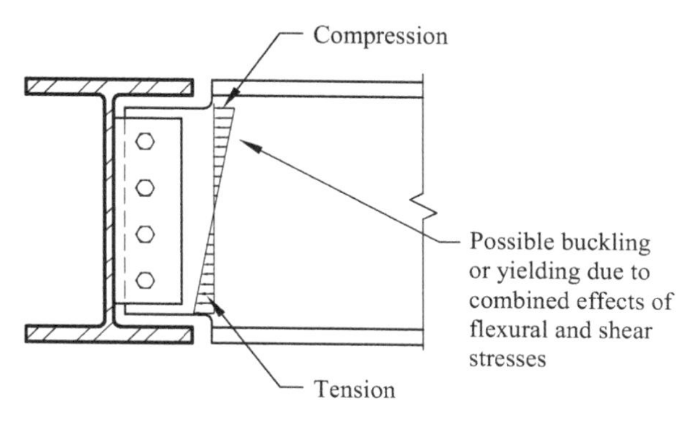

Figure 4-15.(c) Copes required at top and bottom due to some girder depth: Section.

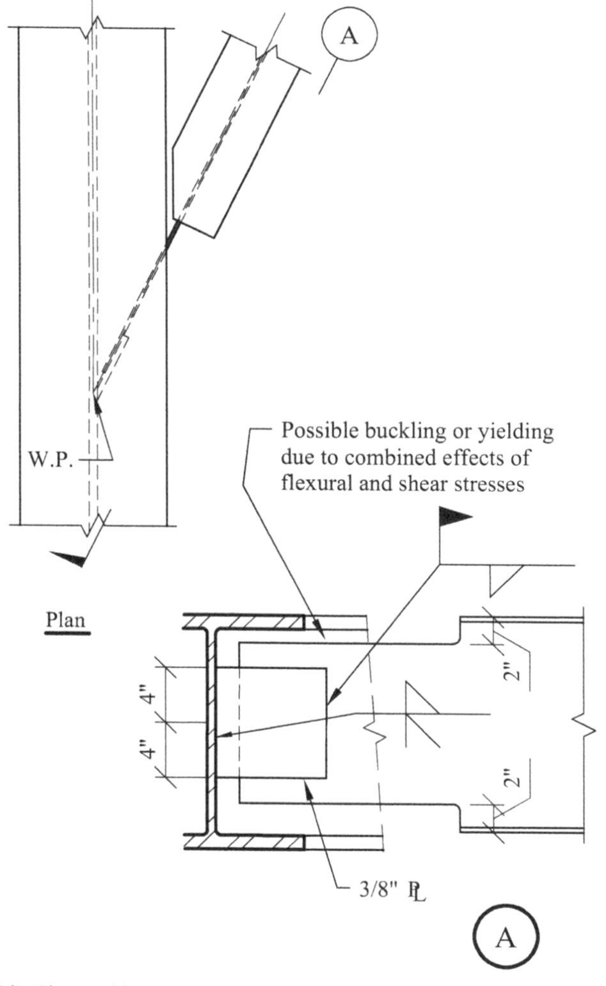

Figure 4-15.(d) Skewed beam to girder connection.

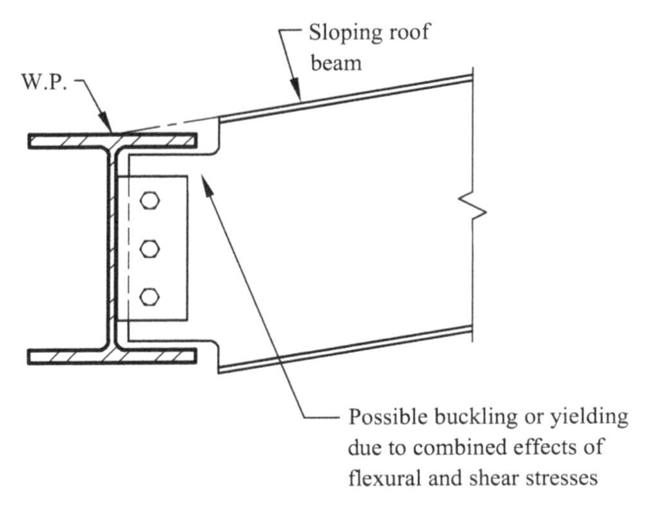

Figure 4-15.(e) Sloping beam to girder connection.

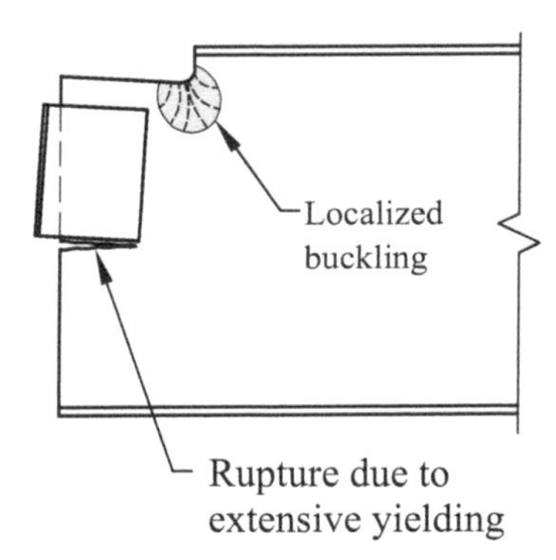

Figure 4-15.(f) Shallow depth shear plate: Block shear and shear buckling.

References

AISC (American Institute of Steel Construction). 2005. *Seismic design provisions for structural steel buildings*. ANSI/AISC 341 Seismic. Chicago: AISC.

AISC. 2017. *Steel construction manual*. 15th ed. Chicago.

Dowswell, B. 2018. "Manualwise: Designing beam copes." In *Modern Steel Construction*. 15th ed. Chicago: AISC, 16–20.

Johnston, B. V. 1976. *Guide to stability design criteria for metal structures structural stability research council*. Hoboken, NJ: Wiley.

Marianian, P. 2010. *Reducing brittle and fatigue failures in steel structures*. Reston, VA: ASCE.

Masabuchi, K. 1980. *Analysis of welded structures*. Oxford, UK: MIT, Pergamon Press.

Timoshenko, S. P. 1955. *Strength of materials. Part 1*. 3rd ed. New York: D. Van Nostrand.

CHAPTER 5

Recommendations

5.1 MEASURES TO MINIMIZE RISK OF FAILURES DURING SERVICE

(a) *Recognizing eccentric conditions*: Chapter 1 has given a few examples in which failures have occurred during service, which were at least partly because of instability conditions. As discussed in previous chapters, there are many issues that can set up instability conditions. These include those listed in Section 2.8 from Wood (1958) and Johnston (1976, pp. 427–428). A persistent issue is often the lack of recognition of eccentric conditions because of the finalization of details, for example, illustrated in the partial failure that occurred at a single-story market, Burbank, California, in 1993 [Figure 1-9(a to d)]. The eccentricities that occurred at the cable anchorages were not apparent on the original structural drawings, and thus appeared to have caused significant secondary stresses that led to the partial failure. Other examples are discussed in Section 4.4.

Therefore, it is very important to develop the details on the contract documents to a level that fully determines the eccentricities that need to be considered in the design and not leave it to the shop-drawing stage, where they may not be so readily accounted for.

(b) *Three-dimensional effects and out-of-plane distortions*: Three-dimensional effects, in many cases, need to be considered where members and/or their connections are expected to yield. Out-of-plane distortions, in combination with forces acting in-plane, are often ignored, which may cause instability. This is particularly true during seismic events when both in-plane and out-of-plane forces can simultaneously occur. Figure 5-1 illustrates possible behavior associated with a concentric brace frame. The frame is subject to in-plane forces, causing compression in the brace along with out-of-plane forces and movements, as shown in Figure 5-1(b, c). The out-of-plane forces and movements can significantly increase distortions, resulting in an increased instability. Even when the brace is in tension, as shown in Figure 5-2(a) with reference to Figure 5-2(b), compression can occur in the gusset plate. Along with out-of-plane forces and movements, there is a potential

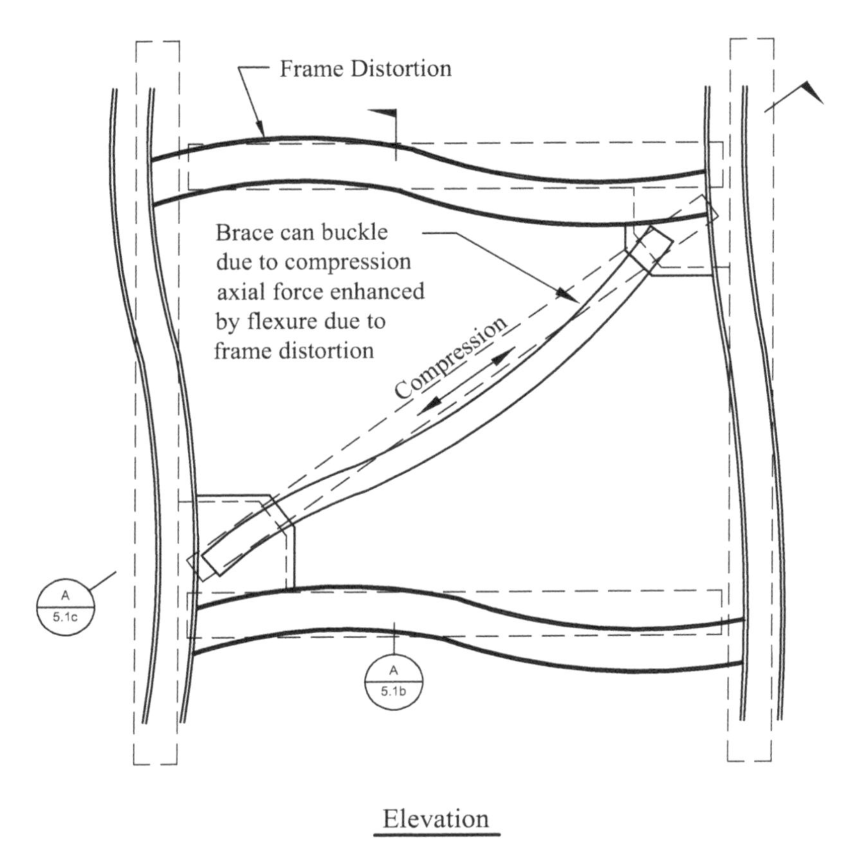

Figure 5-1.(a) Brace frame action: Brace in compression.

for instability to occur. Similar conditions occur with other systems such as moment frames, eccentric brace frames, and buckling restrained braced frames.

(c) *Changes because of temperature and shrinkage:* Thermal movements and changes because of long-term shrinkage and creep of concrete affecting steel members need to be considered. An example is shown in Figure 5-3(a), where the stability of an upturned girder can be affected by thermal movements. Another example is where exposed beams, such as track beams supported by outriggers, are subjected to significant stresses because of restraint to thermal changes, as shown in Figure 5-3(b, c). The restraint to thermal changes, causing compressive forces, can enhance the potential for instability. Furthermore, variations in temperature between the top and bottom surfaces of members can induce secondary moments, enhancing compression forces and possibly causing unintended instability. The

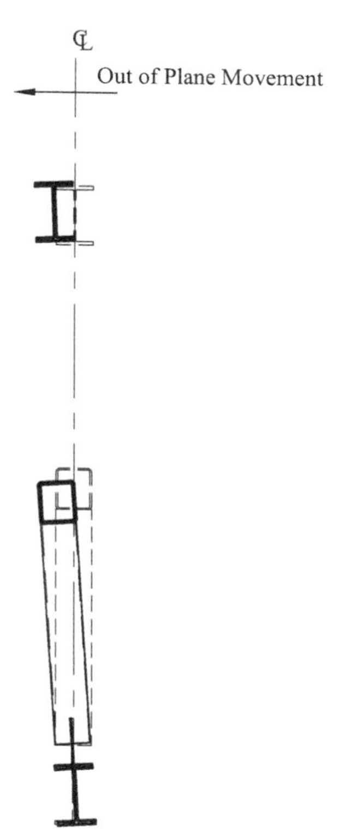

Figure 5-1.(b) Section A–Brace frame action: Brace in compression.

introduction of expansion joints may be necessary to control the stresses and consequentially reduce the potential for instability.

(d) *Bracing for strength and stiffness:* The importance that bracing provides in both sufficient strength and stiffness to prevent instability, as discussed in previous chapters, is essential. Simple testing of conditions, such as shown in Figure 5-4(a), can first establish (on an approximate basis) whether problems occur. Figure 5-4(b) indicates a simple model to test the stiffness of a bent frame for an axially loaded column. Figure 5-4(c) shows the stiffness test to ensure stability for the top flange of a series of girders (in this case, three are shown).

(e) *Redundancy:* Building codes typically do not address considerations on providing redundancy to the structure such that where a failure of any component of a structure does not lead to partial, or at worst, global collapse. Furthermore, in the opinion of the author, current practice is such that designs are invariably carried out to meet minimum code requirements, only such that they lack consideration of the consequences of failures of

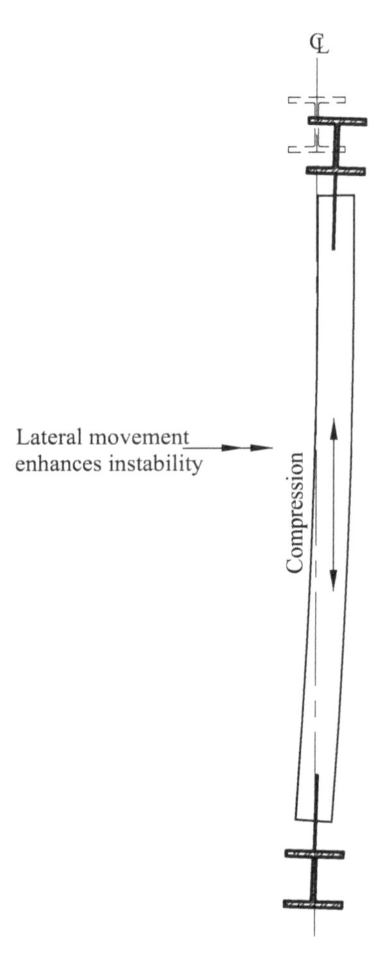

Figure 5-1.(c) Section A–Brace frame action: Brace in compression.

components, including the potential for instability, which could lead to partial or global collapse. There should be a strong initiative to encourage redundancy considerations, particularly for major structures.

(f) *Independent check/peer review:* The importance of carrying out independent checks on the design and drawings, typically by a peer reviewer, when done thoroughly by experienced structural or civil engineers, cannot be overstated. This is particularly applicable for major and complex structures.

(g) *Further research and development:* As discussed in Chapter 2, the subject of buckling has a long history, considering its many forms, including columns, building frames, local buckling, and application to various shapes, including wide flanges, tubes, channels, angle plates, and so on. Yet, in the opinion of

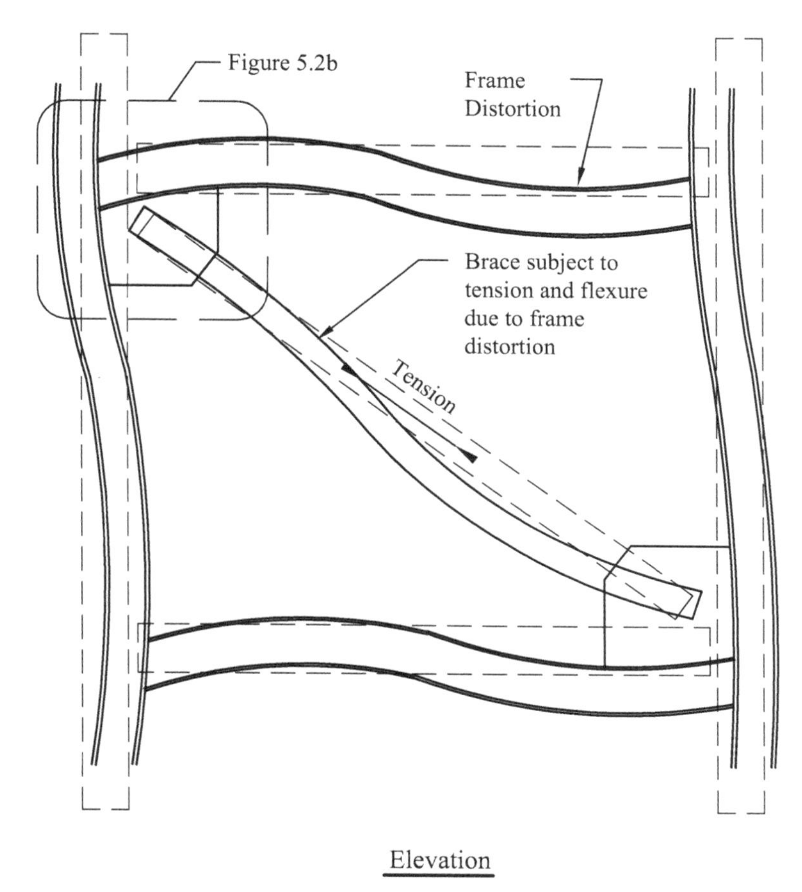

Figure 5-2.(a) Brace frame action brace in tension.

the writer, further research and development appears necessary with regard to local buckling beyond yielding. The local buckling postyield behavior of member components (e.g., flanges and web) needs to be well understood. The findings from such research will need to indicate potential postbuckling behavior such as that illustrated in Figure 2-31(a to c). Appropriate means of controlling postyield local buckling, if expected during a major seismic event, should be developed to reduce damage.

Although buildings may not collapse, significant local buckling will be difficult, if not impossible, to repair. Methods need to be developed to reduce damage and/or facilitate means of repair and partial replacement. A possible example with regard to moment frame connections, with the intent to control local buckling subject to further development and testing, is indicated in Figure 5-5(a to c).

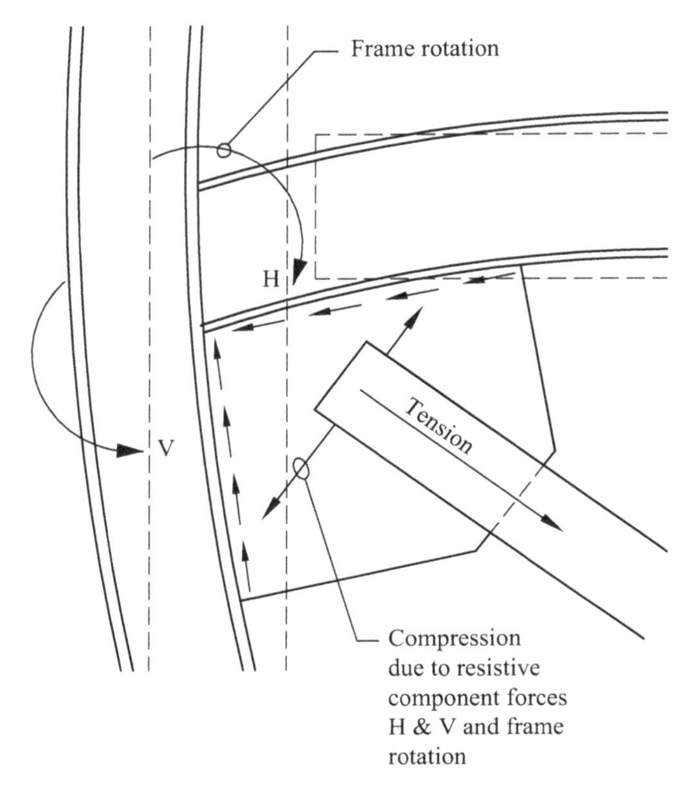

Figure 5-2.(b) Brace frame action: Brace in tension, connection detail from Figure 5-2 (a).

In the opinion of the author, further research appears needed for other systems that involve postyielding of elements such as steel-plated shear walls. As mentioned in Section 2.11 regarding steel-plated shear walls, there remain questions on the behavior of this system including the shortening effects on columns because of gravity loads and overturning forces, along with the complex behavior resulting from higher modes.

Also, it is the opinion of the author that further studies need to be carried out to fully understand the complex behavior of building frames, subjected to dynamic behavior, during seismic events. The frames may have several joints, yielding in a cyclic manner, with the complex dynamic motion affecting their stability. Issues that need to be considered, several of which have been previously discussed, include the following:

(a) Initial out of straightness;

(b) Variable material properties;

(c) Material rolled in material stresses;

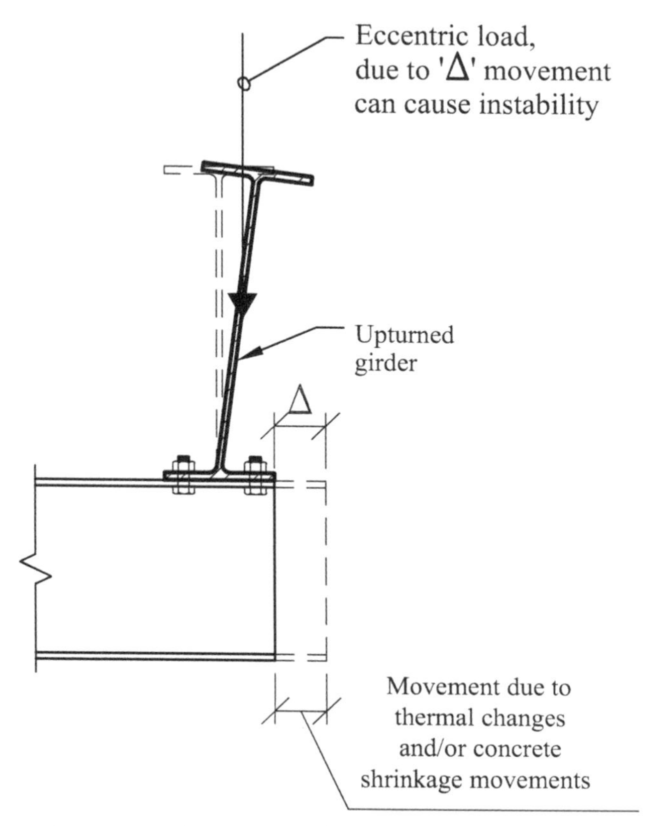

Figure 5-3.(a) Thermal and/or concrete shrinkage movements causing instability.

(d) Residual stresses because of restraint to weld shrinkage;

(e) Column moment magnification;

(f) Out-of-plane motions;

(g) Dynamic cyclic behavior modifying P-delta effects;

(h) Yielding of connections;

(i) Local buckling including panel zones, web, and flanges;

(j) Strain rates affecting material properties Fy and E; and

(k) Fracture potential affected by inadequate fracture toughness, triaxiality, stress and strain concentrations, strain rates, and so on.

Most of the formulas, discussed in Chapter 2, do not properly address instability where large dynamic drifts occur during major earthquakes because they are essential for static applications and are based on assumed boundary conditions. The use of available computer software, incorporating nonlinear finite-element analysis, applying time histories, can be carried out representing most of

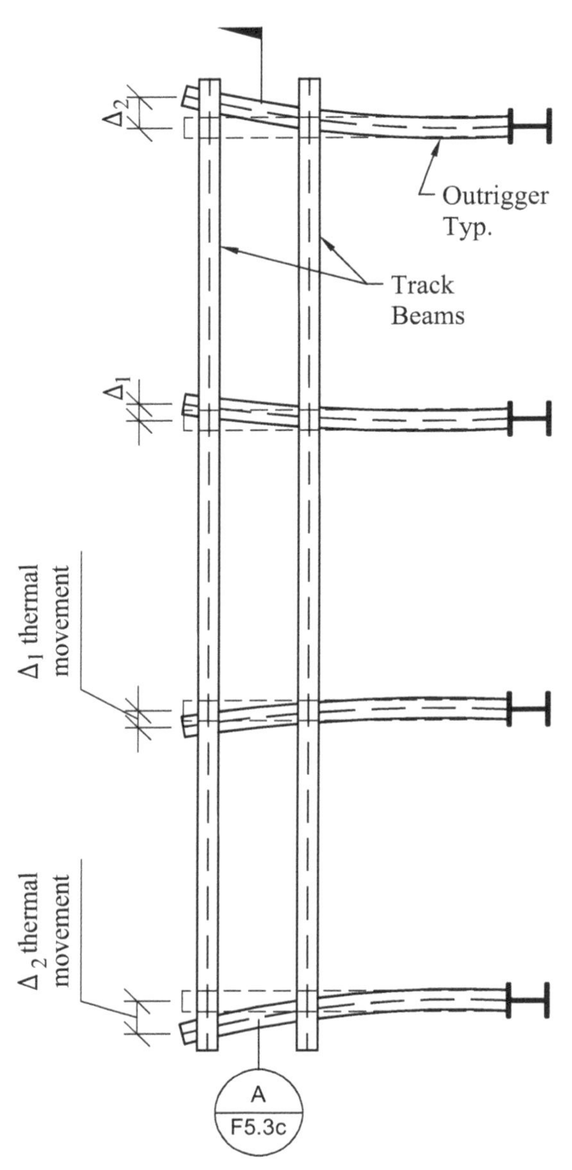

Figure 5-3.(b) Thermal effects on track beams on outriggers: Plan.

the items previously listed. These computer software applications, by performing numerous iterations during the dynamic motions, determine the incremental changes in deformation, from the initial, out of straightness. Buckling modes and the overall stability of frames, with changes in boundary conditions, as the joints transition from being elastic to partially yielding to possible full yielding,

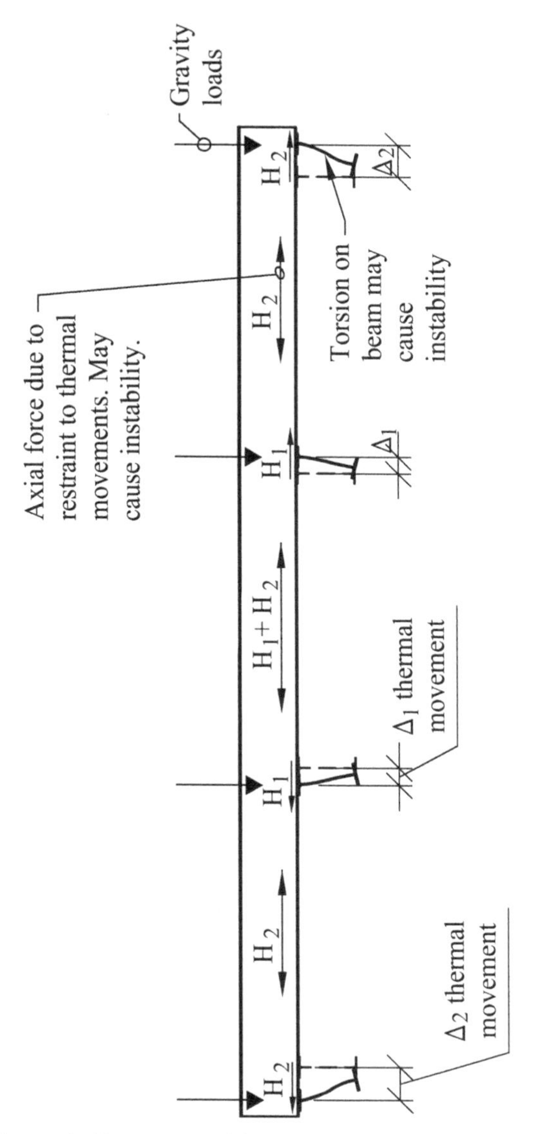

Figure 5-3.(c) Thermal effects on track beams on outrigger–Section A.

are thus accounted for. Items that are not so readily modeled are Item (d) on restraint to weld shrinkage, Item (j) on strain rate, and Item (k) on fracture. It is the author's understanding that software, addressing these issues, is available but would significantly increase the complexities of the analysis. It is the author's recommendation that studies be carried out on various structures, using available software, to better understand the dynamic behavior of structures. This is important

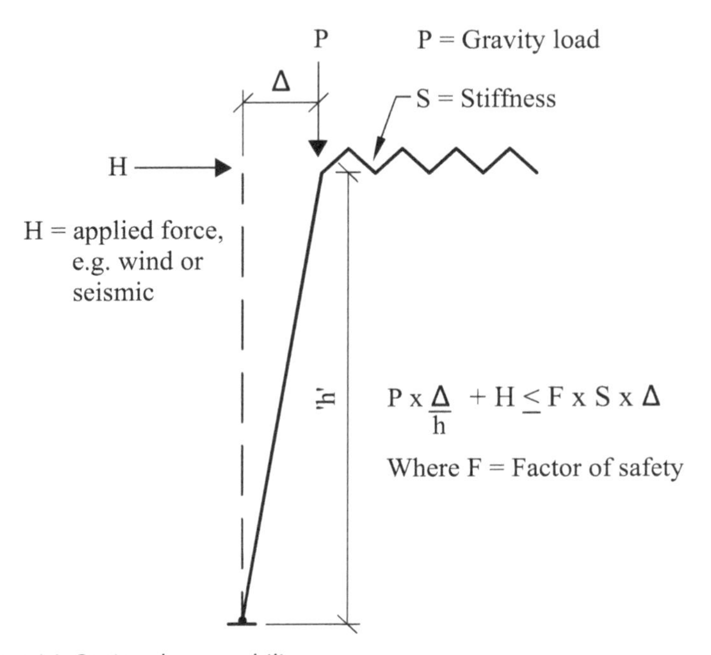

Figure 5-4.(a) Basic column stability.

Note: Stability needs to be considered in both in-plane and out-of-plane directions.

to gain a sound understanding of the structural behavior of stability during major earthquakes and also to assist with establishing procedures for preliminary designs along with gaining some comparisons with static applications. It is hoped that future development will not only lead to a greater sophistication of analysis but also enable more extensive use for the design of structures. The importance of a sound understanding of structural dynamics and that this subject is a significant part of the education of structural engineers is paramount.

5.2 MEASURES TO MINIMIZE THE RISK OF INSTABILITY DURING CONSTRUCTION

(a) *Structural or civil engineer's responsibilities*: A structural or civil engineer, in the opinion of the writer, should have a clear concept of how the structure is to be constructed, considering limitations that may prevail at the site (e.g., access, adjacent properties) Any additional components, such as bracing, to ensure stability during construction should be added to the contract documents, preferably the drawings, indicating that their intent is for temporary construction conditions. Additional requirements such as

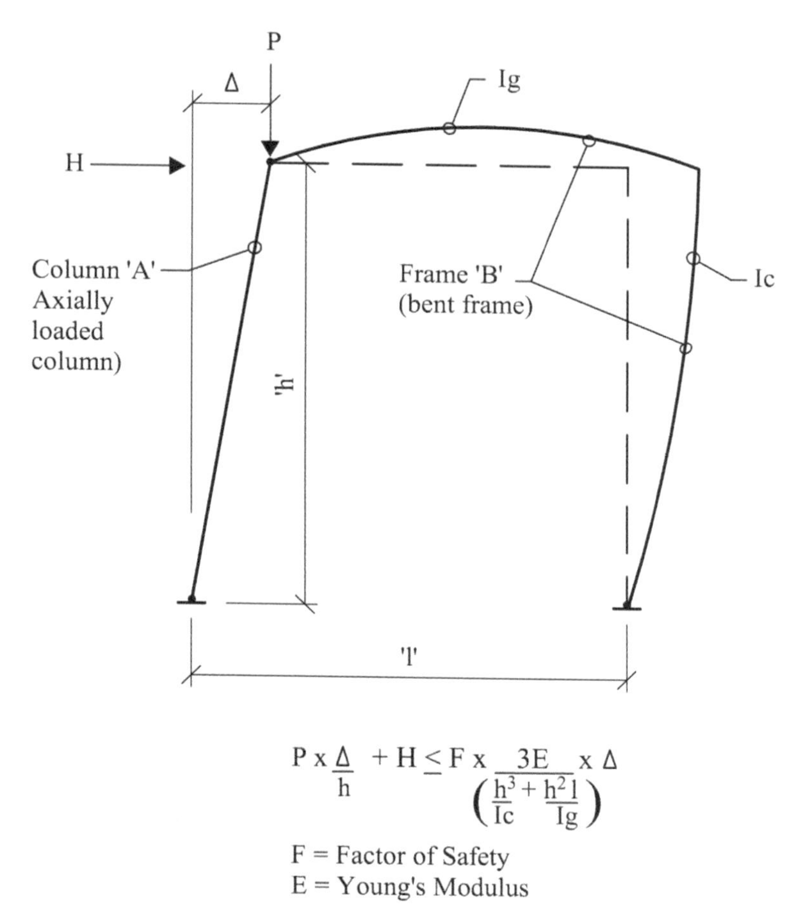

$$P \times \frac{\Delta}{h} + H \leq F \times \frac{3E}{\left(\frac{h^3}{Ic} + \frac{h^2 l}{Ig}\right)} \times \Delta$$

F = Factor of Safety
E = Young's Modulus

Figure 5-4.(b) Example: Column A dependent on the stiffness of frame system "B."

seismic criteria relating to temporary conditions should be clearly indicated on the drawings.

(b) *Process to minimize failures during construction:* Many instances exist during construction in which instability can occur. A few examples are shown in Figures 5-6(a to d). Although these may be regarded as simple cases, they are indicative of a lack of care, which may be enhanced by demands on construction personnel with respect to performance and the need to maintain the schedule.

The key to minimizing the risk of unstable conditions during construction is to have a process in place that includes having personnel with sufficient training and knowledge of the various issues.

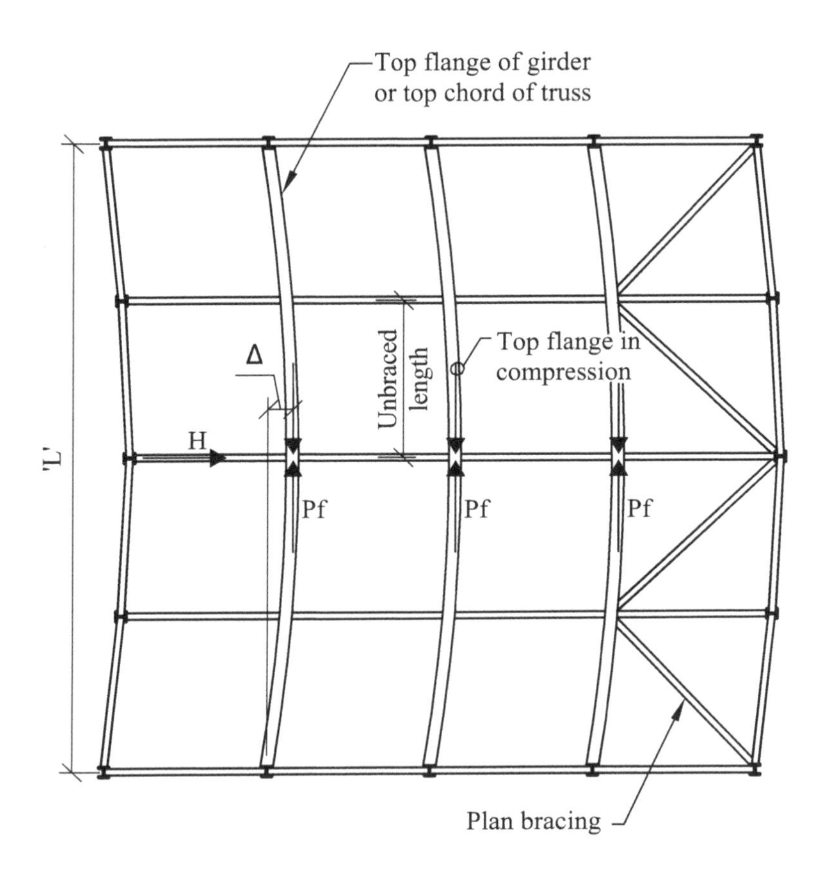

Approximately for stability of the three girders,

$$3 \times Pf \times \frac{\Delta}{L} \times 2 + H \leq F \times S \times \Delta$$

F = Factor of safety
H = applied force, e.g., wind or seismic
Pf = Force
S = Stiffness of plan bracing = P/δ

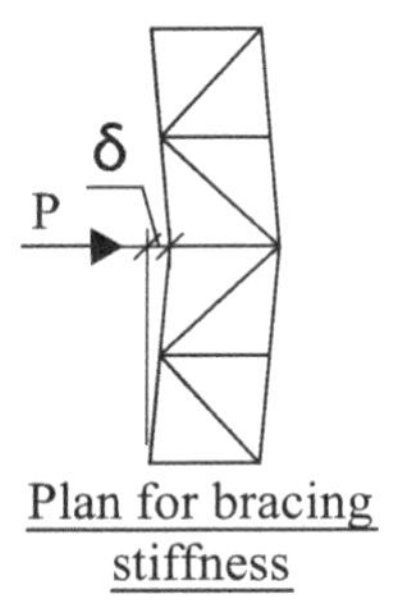

Figure 5-4.(c) Plan at the top flange of girders or top chord of trusses.

The Westgate Bridge disaster, discussed in Chapter 1, is a major example in which poor communication and a lack of proper supervision were cited as contributions to the collapse of the bridge. The process needs to include structural or civil engineer(s), retained specifically for temporary erection and construction. This should be clearly called for on the contract

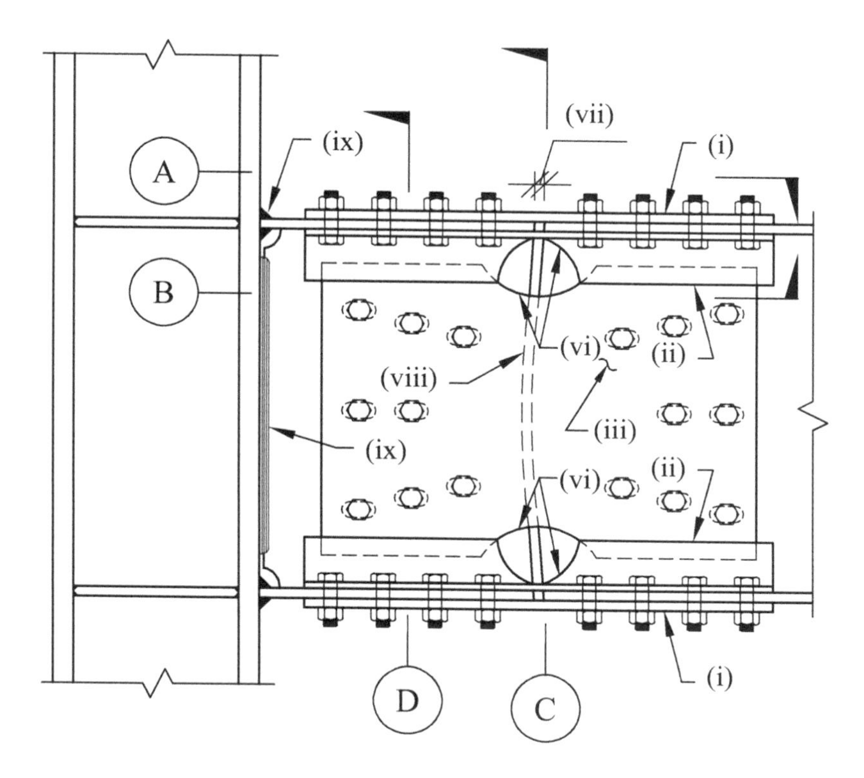

(i) Stl PL GR50
(ii) Stl angle or welded plates to form an angle
(iii) Web PL ea. side
(iv) Elongated hole in plate (i)
(v) Reduced section in angle (ii)
(vi) Curved profile in web PL (iii) and angle (ii)
(vii) Calculated gap
(viii) Profile cut
(ix) Elongated fillet weld both sides (shop welded)

Figure 5-5.(a) Possible replacement moment connection that also controls local buckling.

Note: (i) = Stl PL GR50, (ii) = Stl angle or welded plates to form an angle, (iii) = web PL ea. side, (iv) = elongated hole in plate (i), (v) = reduced section in angle (ii), (vi) = Curved profile in web PL (iii) and angle (ii), (vii) = calculated gap, (viii) = profile cut, (ix) = elongated fillet weld both sides (shop welded).

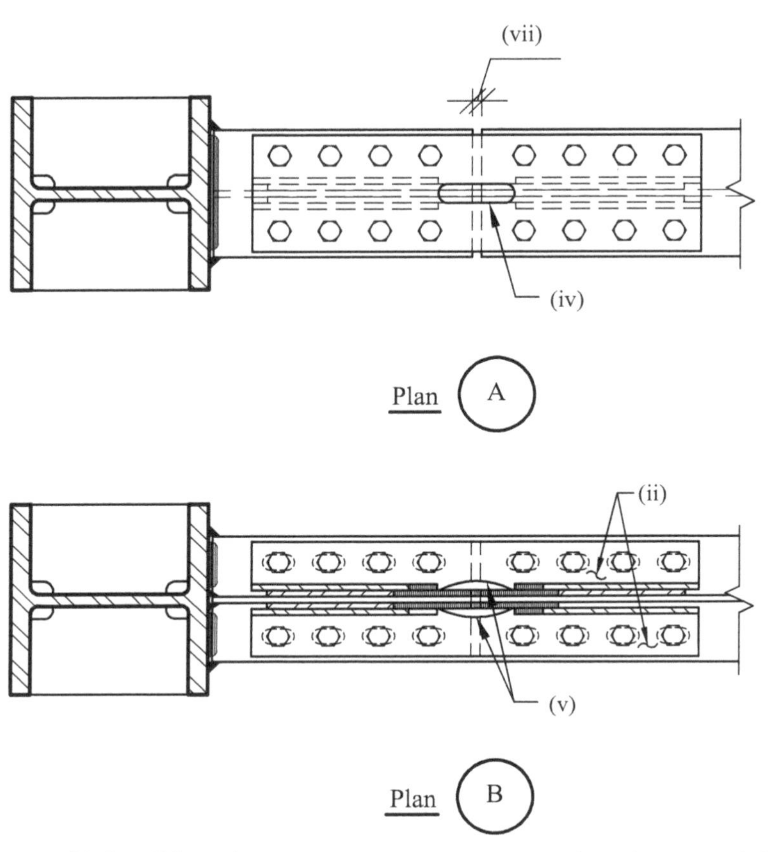

Figure 5-5.(b) Possible replacement moment connection that also controls local buckling: Sections A and B.

documents by the Structural or Civil Engineer of Record. Information that the project team should require for projects for items such as temporary shoring, includes but is not limited to, the following:

(i) Applicable codes and standards and specifications;

(ii) Special criteria, for example, special seismic and/or wind conditions if greater than applicable standards;

(iii) Clearly stated procedures;

(iv) Clearly stated sequencing;

(v) Protocols on communications;

(vi) Testing and inspection requirements; and

(vii) Where deemed necessary, special testing and inspection protocols.

With regard to Item (i), the contractor needs to be aware of all codes, standards, and specifications, and thus, have personnel fully cognizant of the safety requirements of all relevant documents.

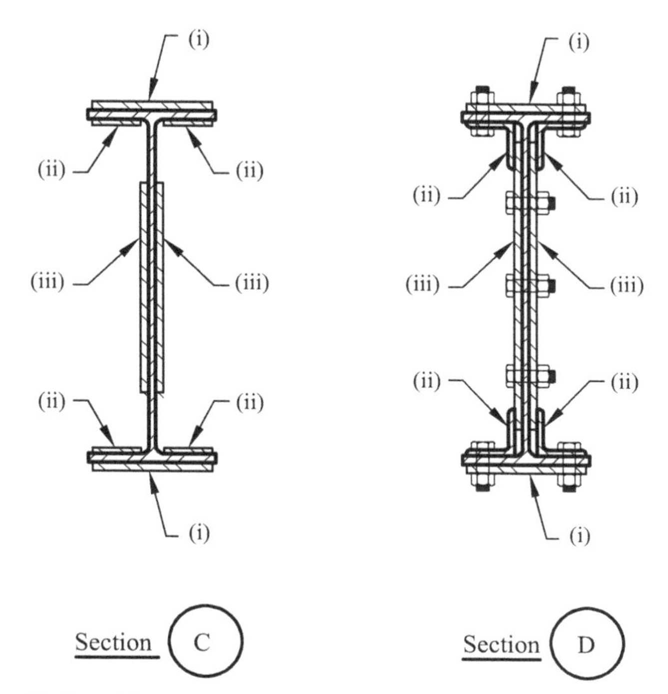

Figure 5-5.(c) Possible replacement moment connection that also controls local buckling: Sections C and D.

With regard to Item (ii), it can be important to know the anticipated duration of the temporary erection because it relates to exposure to risk from a seismic event and/or wind event. Obviously, the longer the duration, the greater the risk that a moderate to major earthquake event may occur. Similarly, the longer the duration, the greater the risk of a moderate or major wind event, along with a consideration of known seasonal weather variations and potentials for storms.

With regard to Items (ii) to (iv), it is important that the procedures and sequencing should be developed by the contractor and subcontractor(s) and their retained structural or civil engineer(s). This should include drawings, calculations, and specifications. It is very important that the structural or civil engineer for the contractor is fully cognizant of the design of the structure and has all the relevant information to develop the temporary erection documents, including relevant shop drawings. Additional consultants may be required. For example, a welding consultant may be required if there is field welding that may cause distortions and a high buildup of residual stresses and strains. As previously discussed in Section 4.3, the distortions from welding may increase the potential for instability. The temporary erection/construction documents should be reviewed by the Structural Engineer of Record to ensure that the contactor's structural or civil engineer has

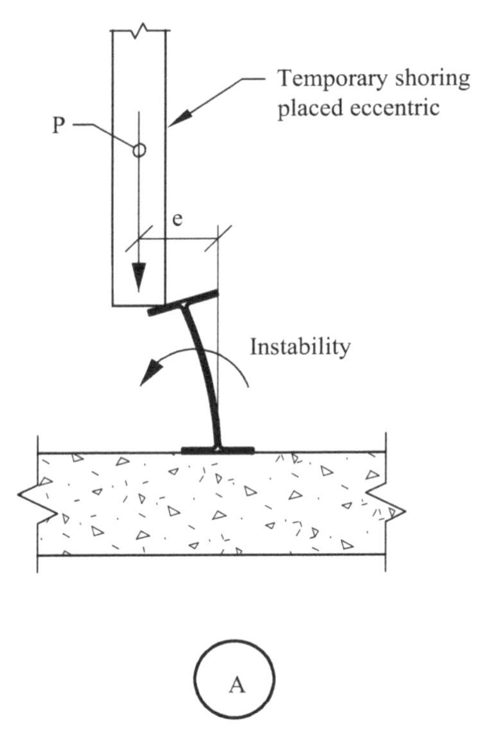

Figure 5-6.(a) Eccentric temporary post.

fully comprehended the building's structural components associated with the temporary erection scope of work. For major/complex structures, an independent peer review may be considered appropriate.

With regard to Item (v), a clear identification of key personnel and correspondence procedures needs to be established early in the process. If several key subcontractors are involved, it is important that the contractor fully coordinates related work and the respective responsibilities between these trades. Misunderstandings are often part of the causes of failures. Preconstruction meetings to include the contractor, subcontractor(s), retained structural or civil engineer, Structural Engineer of Record, inspectors, and other interested parties need to be carried out to establish an understanding of the issues, procedures, sequencing, testing and inspection requirements, protocols, and so on.

With regard to Items (vi) and (vii), a testing and inspection plan corresponding to the procedures and sequencing should be established. For example, because there may be several stages during erection, there may be a need to recheck connections (bolts and welds) that were

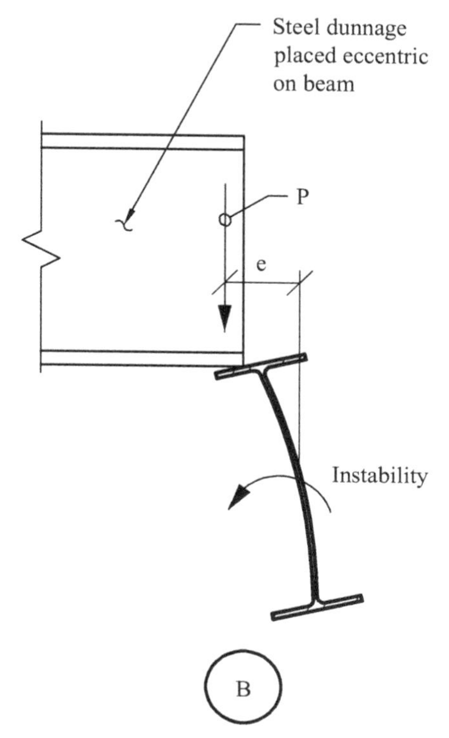

Figure 5-6.(b) Eccentrically placed steel dunnage.

previously tested and inspected. On-site measurement of deformations may also need to be carried out to monitor movements. Although over the decades computer modeling and analysis has been improved, as-built construction may involve additional complex stress conditions because of fit-up, restraint to weld shrinkage stresses, inadvertent eccentric loading conditions, thermal changes, and concrete shrinkage movements and/or additional stresses because of restraint to concrete shrinkage, and so on. The addition of sensors, such as strain gauges or other devices, to assess stress conditions in critical members can assist in monitoring structures during construction and beyond. An improvement in the process of how as-built data information is incorporated into computer structural models, such that costs are reduced, may help toward field sensors, being more financially acceptable. However, an assessment of the structural behavior still needs to be made by knowledgeable and experienced structural engineers, often with the use of hand check calculations supplemental to computer software analysis and calculations, to ensure that conditions are understood.

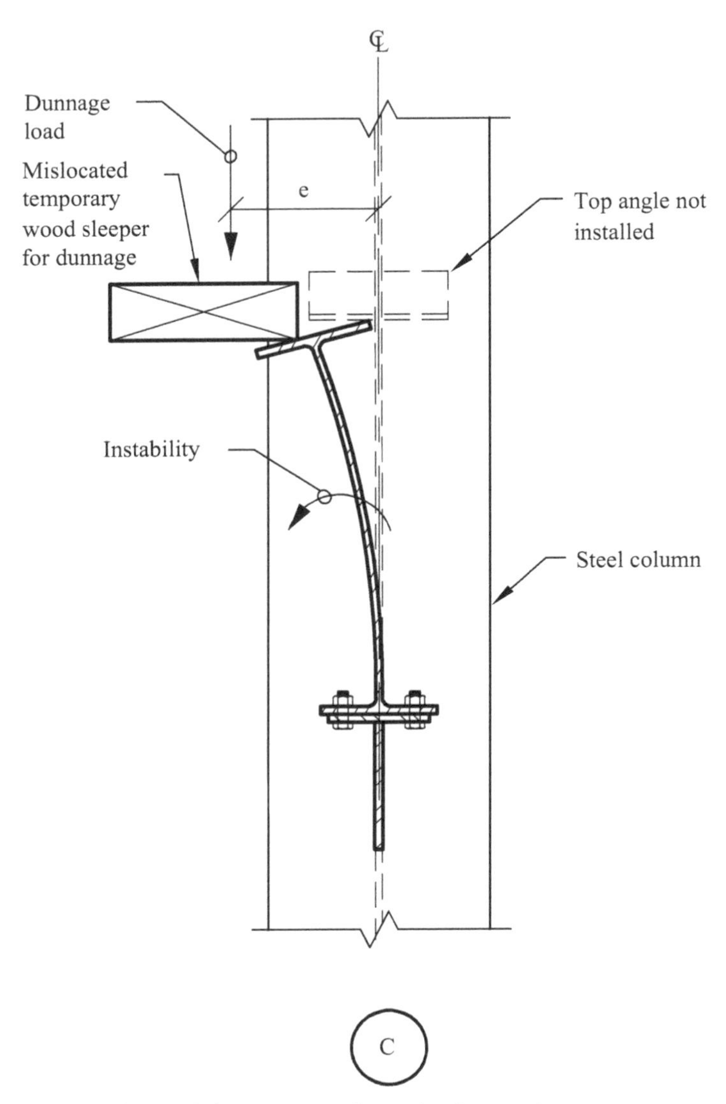

Figure 5-6.(c) Mislocated dunnage; steel erection incomplete.

(c) *Liability issues:* One of the problems that still persists in the structural and civil engineering profession, as it relates to the construction industry, is liability. Structural and civil engineers may see their liability as being only to design the final structure and not wish to be involved in the contractor's means and methods. Contractors often wish to keep their costs to a minimum and see construction problems that may arise to be design issues. Additional costs for the services of structural or civil engineer(s) along with testing and inspection, which may be defined

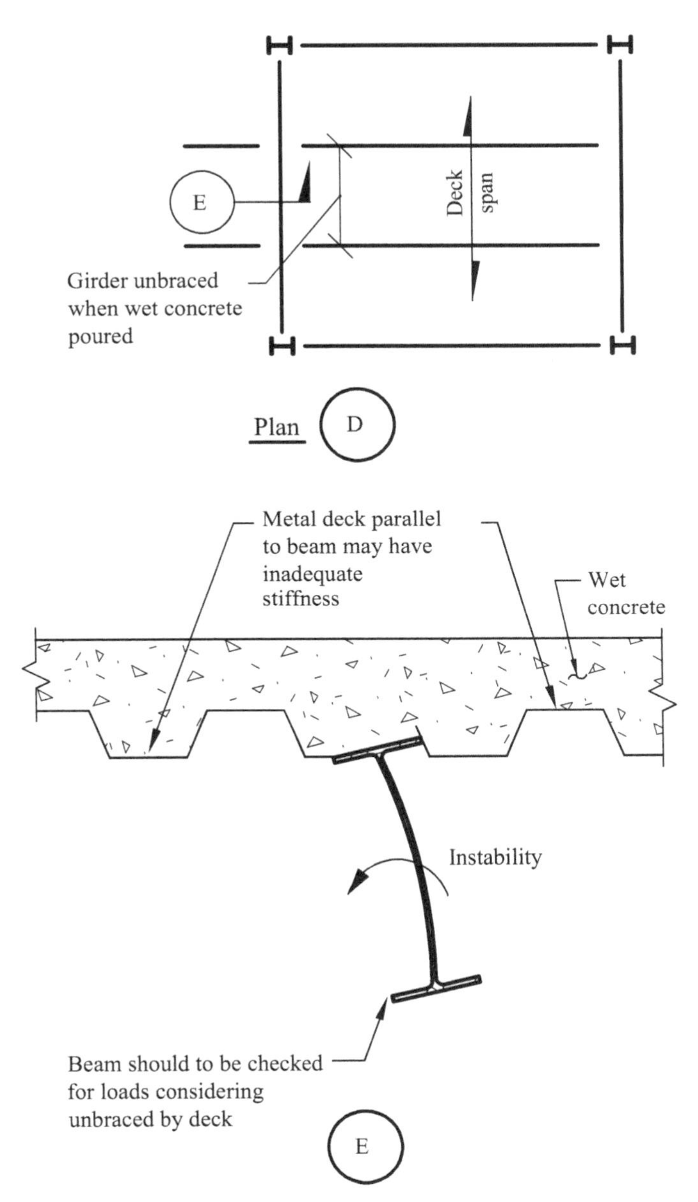

Figure 5-6.(d) Possible instability condition during wet concrete placement.

in construction documents, are often questioned by owners as being beyond code and thus tend to be resisted. This is an ongoing struggle that has persisted forever with unfortunately continued reoccurrence of failures, including those associated with instability, which, of course, is the subject of this publication.

5.3 CONCLUDING REMARKS

As mentioned in the Preface, this publication is intended to give an overview of the subject and associated issues of the potential instability of steel structures and hopefully improve its understanding. The aim is to help avoid the reoccurrence of failures of structures during construction and service that occurred because of a lack of appropriate care with respect to stability.

In the opinion of the author, changes in the design and construction contract process need to be made to effect improvements, particularly for complex structures. These are mostly discussed in Section 5.2. Furthermore, the importance of the assessment/design of structures, using inelastic buckling analysis procedures, is discussed in Section 2.13.

In general, greater cooperation between the design and construction teams is necessary. The importance of engineering design and construction issues, being clearly communicated and understood by clients, is essential as part of the responsibility for their cost of ownership.

Furthermore, it is very important that the education of structural engineers covers in depth the historical development of the extensive methods of evaluating instability so that they can have a thorough understanding of the subject and not merely adhere to prescriptive code requirements. The pioneering work by many researchers during the eighteenth, nineteenth, and twentieth centuries, as described in Chapter 2, led to an invaluable wealth of understanding on the subject of instability. The value of their great work cannot be understated.

This publication has been carried out for ASCE's Forensic Engineering Division's Committee for Practice to Reduce Failures (CPRF). The committee, at the time of writing, comprises fewer than 10 volunteer members, endeavors, by way of publications such as this one, to bring to the attention of practicing engineers' issues that tend to cause failure and provide recommendations to reduce failures. Examples are available from the past in which appropriate actions have been carried out in response to failures, such as the 1973 Merrison Committee's report following the 1970 Milford Haven Bridge Collapse in Wales. However, our committee members, each having their own expertise, are aware that, in some cases, a repetition of similar types of failures still occurs, reflecting a failure to learn from past errors and misunderstandings.

In consideration of the aforementioned, it is important that when issues are brought to the attention of the engineering community, appropriate engineering action is taken through and acted on by code and/or other authorities. As professionals, we must stay true to the basic principles of engineering and continually pursue, by way of research, the understanding of complex issues.

The long history on the study of instability, commencing at least as early as the eighteenth century, should continue. This should include the complex behavior of building frames, subjected to seismic events, where some have joints that have yielded, along with the postyield buckling of joints such as moment connections. Further research should include other lateral resisting systems, such as steel-plated shear walls, along with a further understanding of the effects of welding on global and local buckling and nonlinear buckling.

References

Johnston, B. V. 1976. *Guide to stability design criteria for metal structures structural stability research council.* Hoboken, NJ: Wiley.

Wood, R. H. 1958. "The stability of tall buildings." In *Proc. Inst. Civil Eng.*, 11 (1): 69–102. Also published as a historical paper in Structures and Buildings. *Proc., Inst. Civil Eng.*, 161 (SB5): 247–258.

Index

Note: Page numbers followed by *f* indicate figures.